T0091825

VARIATIONAL INEQUALITIES WITH APPLICATIONS

For other titles published in this series, go to
www.springer.com/series/5613

Advances in Mechanics and Mathematics

VOLUME 18

Aims and Scope

Mechanics and mathematics have been complementary partners since Newton's time, and the history of science shows much evidence of the beneficial influence of these disciplines on each other. The discipline of mechanics, for this series, includes relevant physical and biological phenomena such as: electromagnetic, thermal, quantum effects, biomechanics, nanomechanics, multiscale modeling, dynamical systems, optimization and control, and computational methods.

Driven by increasingly elaborate modern technological applications, the symbiotic relationship between mathematics and mechanics is continually growing. The increasingly large number of specialist journals has generated a complementarity gap between the partners, and this gap continues to widen. *Advances in Mechanics and Mathematics* is a series dedicated to the publication of the latest developments in the interaction between mechanics and mathematics and intends to bridge the gap by providing interdisciplinary publications in the form of monographs, graduate texts, edited volumes, and a special annual book consisting of invited survey articles.

VARIATIONAL INEQUALITIES WITH APPLICATIONS

A Study of Antiplane Frictional Contact Problems

By

MIRCEA SOFONEA
Université de Perpignan, France

ANDALUZIA MATEI
University of Craiova, Romania

Mircea Sofonea
Laboratoire LAMPS
Université de Perpignan
66 860 Perpignan Cedex
France
sofonea@univ-perp.fr

Andaluzia Matei
Department of Mathematics
University of Craiova
200585 Craiova
Romania
andaluziamatei2000@yahoo.com

Series Editors:

David Y. Gao
Department of Mathematics
Virginia Tech
Blacksburg, VA 24061
gao@vt.edu

Ray W. Ogden
Department of Mathematics
University of Glasgow
Glasgow, Scotland, UK
rwo@maths.gla.ac.uk

ISSN: 1571-8689
ISBN: 978-0-387-87459-3   e-ISBN: 978-0-387-87460-9
DOI: 10.1007/978-0-387-87460-9

Library of Congress Control Number: 2008944205

Mathematics Subject Classification (2000): 58E35, 58E50, 49J40, 49J45, 74M10, 74M15, 74G25, 74G30, 74M99

Printed on acid-free paper

springer.com

To my teachers Nicolae Cristescu, Caius Iacob,
and Gheorghe Popescu, with gratitude
(Mircea Sofonea)

To Claudiu, Alexandru, and Robert, with love
(Andaluzia Matei)

Series Preface

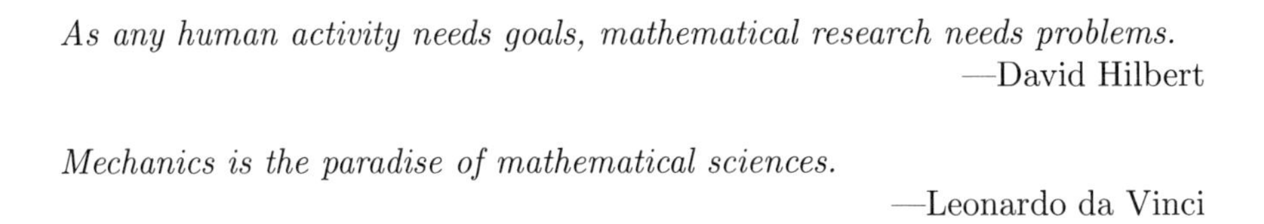

As any human activity needs goals, mathematical research needs problems.
—David Hilbert

Mechanics is the paradise of mathematical sciences.
—Leonardo da Vinci

Mechanics and mathematics have been complementary partners since Newton's time, and the history of science shows much evidence of the beneficial influence of these disciplines on each other. Driven by increasingly elaborate modern technological applications, the symbiotic relationship between mathematics and mechanics is continually growing. However, the increasingly large number of specialist journals has generated a duality gap between the partners, and this gap is growing wider.

Advances in Mechanics and Mathematics (AMMA) is intended to bridge the gap by providing multidisciplinary publications that fall into the two following complementary categories:

1. An annual book dedicated to the latest developments in mechanics and mathematics;
2. Monographs, advanced textbooks, handbooks, edited volumes, and selected conference proceedings.

The AMMA annual book publishes invited and contributed comprehensive research and survey articles within the broad area of modern mechanics and applied mathematics. The discipline of mechanics, for this series, includes relevant physical and biological phenomena such as: electromagnetic, thermal, and quantum effects, biomechanics, nanomechanics, multiscale modeling, dynamical systems, optimization and control, and computation methods. Especially encouraged are articles on mathematical and computational models and methods based on mechanics and their interactions with other fields. All contributions will be reviewed so as to guarantee the highest possible scientific standards. Each chapter will reflect the most recent achievements in the area. The coverage should be conceptual, concentrating on the methodological thinking that will allow the nonspecialist reader to understand it. Discussion of possible future research directions in the area is welcome.

Thus, the annual volumes will provide a continuous documentation of the most recent developments in these active and important interdisciplinary fields. Chapters published in this series could form bases from which possible AMMA monographs or advanced textbooks could be developed.

Volumes published in the second category contain review/research contributions covering various aspects of the topic. Together these will provide an overview of the state-of-the-art in the respective field, extending from an introduction to the subject right up to the frontiers of contemporary research. Certain multidisciplinary topics, such as duality, complementarity, and symmetry in mechanics, mathematics, and physics are of particular interest.

The *Advances in Mechanics and Mathematics* series is directed to all scientists and mathematicians, including advanced students (at the doctoral and postdoctoral levels) at universities and in industry who are interested in mechanics and applied mathematics.

David Y. Gao
Ray W. Ogden

Preface

The theory of variational inequalities plays an important role in the study of both the qualitative and numerical analysis of nonlinear boundary value problems arising in mechanics, physics, and engineering science. For this reason, the mathematical literature dedicated to this field is extensive, and the progress made in the past four decades is impressive. A part of this progress was motivated by new models arising in contact mechanics. At the heart of this theory is the intrinsic inclusion of free boundaries in an elegant mathematical formulation.

Contact between deformable bodies abounds in industry and everyday life. Because of the industrial importance of the physical processes that take place during contact, a considerable effort has been made in their modeling, analysis, numerical analysis and numerical simulations, and, as a result, the mathematical theory of contact mechanics has made impressive progress recently. Owing to their inherent complexity, contact phenomena lead to mathematical models expressed in terms of strongly nonlinear evolutionary problems.

Antiplane shear deformations are one of the simplest classes of deformations that solids can undergo: in antiplane shear (or longitudinal shear) of a cylindrical body, the displacement is parallel to the generators of the cylinder and is independent of the axial coordinate. For this reason, the antiplane problems play a useful role as pilot problems, allowing for various aspects of solutions in solid mechanics to be examined in a particularly simple setting. In recent years, considerable attention has been paid to the analysis of such kinds of problems.

The purpose of this book is to introduce to the reader the theory of variational inequalities with emphasis on the study of contact mechanics and, more specifically, with emphasis on the study of antiplane frictional contact problems. The contents cover both abstract results in the study of variational inequalities as well as the study of specific antiplane frictional contact problems. This includes their modeling and variational analysis. Our intention is to illustrate the cross-fertilization between modeling and applications on the one hand, and nonlinear mathematical analysis on the other hand.

Thus, within the particular setting of antiplane shear, we show how new and nonstandard models in contact mechanics lead to new types of variational inequalities and, conversely, we show how the abstract results on variational inequalities can be applied to prove the unique solvability of the corresponding contact problems. In writing this book, our aim was also to draw the attention of the applied mathematics community to interesting two-dimensional models arising in solid mechanics, involving a single nonlinear partial differential equation that has the virtue of relative mathematical simplicity without loss of essential physical relevance.

Our book, divided into four parts with 11 chapters, is intended as a unified and readily accessible source for mathematicians, applied mathematicians, engineers, and scientists, as well as advanced graduate students. It is organized with two different aims, so that readers who are not interested in modeling and applications can skip Parts III and IV and will find an elementary introduction to the theory of variational inequalities in Part II of the book; alternatively, readers who are interested in modeling and applications will find in Parts III and IV the mechanical models that lead to the various classes of variational inequalities presented in Part II of the book.

A brief description of the parts of the book follows.

Part I is devoted to the basic notation and results that are fundamental to the developments later in this book. We review the background on functional analysis and function spaces that we need in the study of variational inequalities. The material presented is standard and can be found in many textbooks and monographs. For this reason, we present only very few details of the proofs.

Part II represents one of the main parts of the book and includes original results. We present various classes of variational inequalities for which we prove existence results and, for some of them, we prove uniqueness, regularity, and convergence results. To this end we use convexity, monotonicity, compactness, time discretization, regularization, and fixed point arguments. Most of the concepts and results presented in this part can be extended to more general variational inequalities involving nonlinear operators on reflexive Banach spaces or to hemivariational inequalities; however, since our aim is to provide an accessible presentation of the theory of variational inequalities with emphasis in the study of antiplane frictional contact problems, we restrict ourselves to the framework of Hilbert spaces, linear operators, and convex analysis, as is sufficient for later development.

The terminology we use in this part of book is the following: if the time derivative of the unknown function u appears in the formulation of a variational inequality (and, therefore, an initial condition for u is needed), we refer to it as an *evolutionary variational inequality.* Otherwise, we refer to it as an *elliptic variational inequality.* If the nondifferentiable convex functional j depends explicitly on u or on its time derivative $\dot{u}$, we refer to the corresponding variational inequality as a *quasivariational inequality.* If both the data and the solution of a variational inequality depend on the time variable that plays the role of a parameter, the corresponding variational inequality is called a *time-dependent variational inequality*. Finally, if an integral term containing the solution or its derivative appears in the formulation of a variational inequality, we refer to it as a *history-dependent variational inequality*. This classification is not strict and is intended to distinguish among the types of variational inequalities used in the mathematical theory of contact mechanics, as it is illustrated in Part IV.

Part III presents preliminary material of contact mechanics that is needed in the rest of the book. We summarize basic notions and equations of mechanics of continua, then we introduce the frictional contact conditions as well as the constitutive laws that are used in the rest of the book. We then specialize the equations and conditions in the context of the antiplane shear and, as an example, we study a displacement-traction problem involving linearly elastic materials. The material presented in this part provides the background for the modeling of the antiplane frictional contact problems studied in Part IV of the book.

Part IV represents the other main part of the book and is partially based on our original research. It deals with the study of static and quasistatic frictional antiplane contact problems. We model the material behavior with isotropic linearly elastic and viscoelastic constitutive laws and, in the case of viscoelastic materials, we consider both short and long memory. Friction is modeled with versions of Coulomb's law in which the friction bound is either a function that does not depend on the process variables or depends on the slip or slip rate. Particular attention is paid to history-dependent frictional problems in which the friction bound depends on the total slip or the total slip rate. For each one of the problems, we provide a variational formulation then we use the abstract results in Part II in order to establish existence and sometimes uniqueness, regularity, and convergence results.

Each of the four parts of the book is divided into several chapters. All the chapters are numbered consecutively. Mathematical relations (equalities, inequalities, and inclusions) are numbered by chapter and their order of occurrence. For example, (4.3) is the third numbered mathematical relation in Chapter 4. Definitions, problems, theorems, propositions, lemmas, and corollaries are numbered consecutively within each chapter. For example, in Chapter 9, Problem 9.5 is followed by Theorem 9.6.

Each part ends with a section in which we present bibliographical comments. We provide references for the principal results presented, as well as information on important topics related to but not included in the body of the text. The list of the references at the end of the book includes only papers or books that are closely related to the subjects treated in this monograph.

This book is a result of cooperation between the authors during the past several years and was partially supported by the Integrated Action France-Romania *Brâncuşi* No. 06080RF/03. Part of the material is based on the Ph.D. thesis of the second author as well as on our joint work with several collaborators to whom we express our thanks. We especially thank Weimin Han,

Constantin Niculescu, Vicenţiu Rădulescu, Meir Shillor, and Juan M. Viaño for our beneficial cooperation and for their constant support. We extend our gratitude to David Y. Gao for inviting us to make the contribution in the Springer book series on *Advances in Mechanics and Mathematics* (AMMA). Finally, we thank the unknown referees for their valuable suggestions, which improved the final form of the book.

Perpignan, France *Mircea Sofonea*
Craiova, Romania *Andaluzia Matei*

July 2008

Contents

List of Symbols

Sets

$\mathbb{N}$: the set of positive integers;

$\mathbb{Z}_+$: the set of non-negative integers;

$\mathbb{R}$: the real line;

$\mathbb{R}_+$: the set of non-negative real numbers;

$\mathbb{R}^d$: the d-dimensional Euclidean space;

$\mathbb{S}^d$: the space of second-order symmetric tensors on $\mathbb{R}^d$;

Ω: an open, bounded, connected set in $\mathbb{R}^d$ with a Lipschitz boundary Γ;

Γ: the boundary of the domain Ω, which is decomposed as $\Gamma = \Gamma_1 \cup \Gamma_2 \cup \Gamma_3$ with Γ_1, Γ_2, and Γ_3 having mutually disjoint interiors;

Γ_1: the part of the boundary where displacement condition is specified; $\operatorname{meas}(\Gamma_1) > 0$ is assumed throughout the book;

Γ_2: the part of the boundary where traction condition is specified;

Γ_3: the part of the boundary where contact takes place;

$[0, T]$: time interval of interest, $T > 0$.

Operators

∇: the gradient operator (pages 14, 29, 149);

Div: the divergence operator (page 130);

div: the divergence operator (page 149);

γ: the trace operator (page 28);

∂_ν: the normal derivative operator (page 151);

$\mathcal{P}_K$: the projection operator onto a set K (page 11);

$\boldsymbol{I}$: the identity operator on $\mathbb{R}^3$ (page 137).

Function spaces

$C^m(\overline{\Omega})$: the space of functions whose derivatives up to and including order m are continuous up to the boundary Γ (page 22);

$C_0^\infty(\Omega)$: the space of infinitely differentiable functions with compact support in Ω (page 22);

$L^p(\Omega)$: the Lebesgue space of p-integrable functions, with the usual modification if $p = \infty$ (page 23);

$W^{k,p}(\Omega)$: the Sobolev space of functions whose weak derivatives of orders less than or equal to k are p-integrable on Ω (page 25);

$H^k(\Omega) \equiv W^{k,2}(\Omega)$ (page 25);

$W_0^{k,p}(\Omega)$: the closure of $C_0^\infty(\Omega)$ in $W^{k,p}(\Omega)$ (page 26);

$H_0^k(\Omega) \equiv W_0^{k,2}(\Omega)$ (page 26);

$V = \{ v \in H^1(\Omega) : v = 0 \text{ a.e. on } \Gamma_1 \}$, with inner product $(u, v)_V = (\nabla u, \nabla v)_{L^2(\Omega)^2}$ (page 152);

X: a Hilbert space with inner product $(\cdot, \cdot)_X$, or a Banach space with norm $\| \cdot \|_X$;

$\mathcal{L}(X, Y)$: the space of linear continuous operators from X to a normed space Y (page 6);

$\mathcal{L}(X) \equiv \mathcal{L}(X, X)$ (page 6);

$X \times Y$: the product of the Hilbert spaces X and Y, with inner product $(\cdot, \cdot)_{X \times Y}$ (page 12);

0_X: the zero element of X;

$C^m([0,T]; X) = \{ v \in C([0,T]; X) : v^{(j)} \in C([0,T]; X),\ j = 1, \ldots, m \}$ (page 33);

$L^p(0,T;X) = \{ v : (0,T) \to X \text{ measurable: } \|v\|_{L^p(0,T;X)} < \infty \}$ (page 33);

$W^{k,p}(0,T;X) = \{ v \in L^p(0,T;X) : \|v^{(j)}\|_{L^p(0,T;X)} < \infty\ \forall j \le k \}$ (page 35);

$H^k(0,T;X) \equiv W^{k,2}(0,T;X)$ (page 35).

Other symbols

c: a generic positive constant;

$r_+ = \max\{0, r\}$: positive part of r;

$\forall$: for all;

$\exists$: there exist(s);

$\Longrightarrow$: implies;

$\overline{A}$: the closure of the set A;

∂A: the boundary of the set A;

δ_{ij}: the Kronecker delta;

a.e.: almost everywhere;

iff: if and only if;

l.s.c.: lower semicontinuous (page 12);

ψ_K: the indicator function of the set K (page 13);

$\partial\varphi$: the subdifferential of the function φ (page 15);

$\dot{f}$: the time derivative of the function f.

Part I
Background on Functional Analysis

Chapter 1
Preliminaries

This chapter presents preliminary material from functional analysis that will be used in subsequent chapters. Most of the results are stated without proofs, as they are standard and can be found in many references. We start with a review of definitions and properties of linear normed spaces and Banach spaces, including results on duality and weak convergence. We then recall some properties of the Hilbert spaces. Finally, we present miscellaneous results that will be applied repeatedly in this book; they include elements of convex analysis, fixed point theorems, and well-known inequalities. All the linear spaces considered in this book including abstract normed spaces, Banach spaces, Hilbert spaces, and various function spaces are assumed to be real spaces. We assume that the reader has some knowledge of linear algebra and general topology.

1.1 Linear Operators on Normed Spaces

The notion of a norm in a general linear space is an extension of the ordinary length of a vector in $\mathbb{R}^2$ or $\mathbb{R}^3$ and is provided by the following definition.

Definition 1.1. Given a linear space X, a *norm* $\|\cdot\|_X$ is a function from X to $\mathbb{R}$ with the following properties.

1. $\|u\|_X \geq 0 \ \forall\, u \in X$, and $\|u\|_X = 0$ iff $u = 0_X$.
2. $\|\alpha\, u\|_X = |\alpha|\, \|u\|_X \ \forall\, u \in X, \forall\, \alpha \in \mathbb{R}$.
3. $\|u + v\|_X \leq \|u\|_X + \|v\|_X \ \forall\, u, v \in X$.

The pair $(X, \|\cdot\|_X)$ is called a *normed space*.

Here and everywhere in this book, 0_X will denote the zero element of X. Also, we will simply say X is a normed space when the definition of the norm is understood from the context.

On a linear space, various norms can be defined. Sometimes, it is desirable to know if two norms are related and, for this reason, we introduce the following definition.

Definition 1.2. Let $\|\cdot\|^{(1)}$ and $\|\cdot\|^{(2)}$ be two norms over a linear space X. The two norms are said to be *equivalent* if there exist two constants c_1, $c_2 > 0$ such that

$$c_1 \, \|u\|^{(1)} \leq \|u\|^{(2)} \leq c_2 \, \|u\|^{(1)} \quad \forall \, u \in X. \tag{1.1}$$

The notion of a seminorm is useful in the study of various nonlinear boundary problems and in error estimates of some numerical approximations.

Definition 1.3. Given a linear space X, a *seminorm* $|\cdot|_X$ is a function from X to $\mathbb{R}$ satisfying the following properties.

1. $|u|_X \geq 0 \; \forall \, u \in X$.
2. $|\alpha \, u|_X = |\alpha| \, |u|_X \; \forall \, u \in X, \, \forall \, \alpha \in \mathbb{R}$.
3. $|u + v|_X \leq |u|_X + |v|_X \; \forall \, u, v \in X$.

It follows from above that a seminorm satisfies the properties of a norm except that $|u|_X = 0$ does not necessarily imply $u = 0_X$.

With a norm at our disposal, we use the quantity $\|u - v\|_X$ to measure the distance between u and v. Consequently, the norm is used to define the bounded sets and the convergence of sequences in the space X.

Definition 1.4. Let $(X, \|\cdot\|_X)$ be a normed space. A subset $A \subset X$ is *bounded* if there exists $M > 0$ such that $\|u\|_X \leq M$ for all $u \in A$. A sequence $\{u_n\} \subset X$ is *bounded* if there exists $M > 0$ such that $\|u_n\|_X \leq M$ for all $n \in \mathbb{N}$ or, equivalently, if $\sup_n \|u_n\|_X < \infty$.

Definition 1.5. Let X be a normed space. A sequence $\{u_n\} \subset X$ is said to converge (strongly) to $u \in X$ if

$$\|u_n - u\|_X \to 0 \quad \text{as } n \to \infty.$$

In this case, u is called the (strong) *limit* of the sequence $\{u_n\}$ and we write

$$u = \lim_{n\to\infty} u_n \quad \text{or} \quad u_n \to u \quad \text{in } X.$$

It is easy to verify that a limit of a sequence, if it exists, is unique. The adjective "strong" is introduced in the previous definition to distinguish this convergence from other types of convergence that will be introduced in the next section. Using (1.1) it is easy to see that, for two equivalent norms, convergence in one norm implies the convergence in the other norm.

The convergence of sequences is used to introduce closed sets and dense sets in a normed space.

Definition 1.6. Let A be a subset of a normed space X. The *closure* $\overline{A}$ of A is the union of A and the set of the limits of all the convergent sequences from A. The set A is said to be *closed* if $\overline{A} = A$ and *dense* if $\overline{A} = X$.

To test the convergence of a sequence without knowing the limiting element, it is usually convenient to refer to the notion of a Cauchy sequence.

Definition 1.7. Let X be a normed space. A sequence $\{u_n\} \subset X$ is called a *Cauchy sequence* if $\|u_m - u_n\|_X \to 0$ as $m, n \to \infty$.

Obviously, a convergent sequence is a Cauchy sequence but in a general infinite dimensional space, a Cauchy sequence may fail to converge. This justifies the following definition.

Definition 1.8. A normed space is said to be *complete* if every Cauchy sequence from the space converges to an element in the space. A complete normed space is called a *Banach space*.

Given two linear spaces X and Y, an *operator* $T : X \to Y$ is a rule that assigns to each element in X a unique element in Y. A real-valued operator defined on a linear space X is called a *functional.* If both X and Y are normed spaces, we can consider the continuity and Lipschitz continuity of the operators.

Definition 1.9. Let $(X, \|\cdot\|_X)$ and $(Y, \|\cdot\|_Y)$ be two normed spaces. An operator $T : X \to Y$ is said to be

1. *continuous* at $u \in X$ if

$$u_n \to u \text{ in } X \Longrightarrow T(u_n) \to T(u) \text{ in } Y;$$

2. *continuous* if it is continuous at each element of the space X;
3. *Lipschitz continuous* if there exists $L_T > 0$ such that

$$\|T(u) - T(v)\|_Y \leq L_T \|u - v\|_X \quad \forall\, u,\, v \in X.$$

Clearly, if T is Lipschitz continuous, then it is a continuous operator, but the converse is not true in general.

We now consider a particular, yet important, type of operators called linear operators.

Definition 1.10. Let X and Y be two linear spaces. An operator $L : X \to Y$ is called *linear* if

$$L(\alpha_1 u_1 + \alpha_2 u_2) = \alpha_1 L(u_1) + \alpha_2 L(u_2) \quad \forall\, u_1, u_2 \in X,\ \alpha_1, \alpha_2 \in \mathbb{R}.$$

For a linear operator L, we usually write $L(v)$ as Lv. For the sake of simplicity, we sometimes write Lv even when L is not linear. A well-known important property of a linear operator is the following.

Theorem 1.11. *Let X and Y be normed spaces and let $L : X \to Y$ be a linear operator. Then L is continuous on X iff there exists $M > 0$ such that*

$$\|Lu\|_Y \le M\|u\|_X \quad \forall\, u \in X.$$

From Theorem 1.11, we conclude that for a linear operator, continuity and Lipschitz continuity are equivalent.

We will use the notation $\mathcal{L}(X,Y)$ for the space of linear continuous operators from a normed space X to another normed space Y. In the special case $Y = X$, we use $\mathcal{L}(X)$ to replace $\mathcal{L}(X,X)$. For $L \in \mathcal{L}(X,Y)$, the quantity

$$\|L\|_{\mathcal{L}(X,Y)} = \sup_{0_X \ne u \in X} \frac{\|Lu\|_Y}{\|u\|_X} \tag{1.2}$$

is called the *operator norm* of L and $L \mapsto \|L\|_{\mathcal{L}(X,Y)}$ defines a norm on the space $\mathcal{L}(X,Y)$. The norm (1.2) enjoys the following compatibility property

$$\|Lu\|_Y \le \|L\|_{\mathcal{L}(X,Y)}\|u\|_X \quad \forall\, u \in X.$$

Moreover, the following result holds.

Theorem 1.12. *Let X be a normed space and let Y be a Banach space. Then $\mathcal{L}(X,Y)$ is a Banach space.*

Later in the book, we will need the concept of compact operators.

Definition 1.13. Let X and Y be two normed spaces and $L : X \to Y$ be a linear operator. The operator L is said to be *compact* if for every bounded sequence $\{u_n\} \subset X$, the sequence $\{Lu_n\} \subset Y$ has a subsequence converging in Y.

The previous definition shows, in other words, that a linear operator $L : X \to Y$ is compact if for each sequence $\{u_n\} \subset X$ that satisfies the inequality $\sup_n \|u_n\|_X < \infty$, we can find a subsequence $\{u_{n_k}\} \subset \{u_n\}$ and an element $y \in Y$ such that $Lu_{n_k} \to y$ in Y. Compact operators are also called *completely continuous* operators.

We now consider an important type of real valued mappings defined on a product of linear spaces.

Definition 1.14. Let X and Y be linear spaces. A mapping $a : X \times Y \to \mathbb{R}$ is called *bilinear form* if it is linear in each variable, that is, for every $u_1,\, u_2,\, u \in X$, $v_1,\, v_2,\, v \in Y$, and $\alpha_1,\, \alpha_2 \in \mathbb{R}$,

$$\begin{aligned} a(\alpha_1 u_1 + \alpha_2 u_2, v) &= \alpha_1\, a(u_1, v) + \alpha_2\, a(u_2, v), \\ a(u, \alpha_1 v_1 + \alpha_2 v_2) &= \alpha_1\, a(u, v_1) + \alpha_2\, a(u, v_2). \end{aligned}$$

In the case $X = Y$, we say that a bilinear form is *symmetric* if

$$a(u,v) = a(v,u) \quad \forall\, u, v \in X.$$

If both X and Y are normed spaces, we can consider the continuity of the bilinear forms.

Definition 1.15. Let $(X, \|\cdot\|_X)$ and $(Y, \|\cdot\|_Y)$ be two normed spaces. A bilinear form $a : X \times Y \to \mathbb{R}$ is said to be *continuous* if there exists a constant $M > 0$ such that

$$|a(u,v)| \leq M \, \|u\|_X \|v\|_Y \quad \forall\, u \in X, \ \forall\, v \in Y.$$

In the case $X = Y$, we say that a bilinear form is X*-elliptic* if there exists a constant $m > 0$ such that

$$a(u,u) \geq m \, \|u\|_X^2 \quad \forall\, u \in X.$$

Bilinear symmetric continuous and X-elliptic forms defined on a Hilbert space X will be used in Part II of this book in the study of variational and quasivariational inequalities.

1.2 Duality and Weak Convergence

For a normed space X, the space $\mathcal{L}(X, \mathbb{R})$ is called the *dual space* of X and is denoted by X'. The elements of X' are *linear continuous functionals* on X. The duality pairing between X' and X is usually denoted by $\ell(u)$ or $\langle u', u \rangle$ for $\ell, u' \in X'$ and $u \in X$. As it follows from (1.2), the norm on X' is given by

$$\|\ell\|_{X'} = \sup_{0_X \neq u \in X} \frac{|\ell(u)|}{\|u\|_X}.$$

Also, by Theorem 1.12 we know that $(X', \|\cdot\|_{X'})$ is a Banach space.

We can now introduce another kind of convergence in a normed space.

Definition 1.16. Let X be a normed space. A sequence $\{u_n\} \subset X$ is said to *converge weakly* to $u \in X$ if for every $\ell \in X'$,

$$\ell(u_n) \to \ell(u) \quad \text{as } n \to \infty.$$

In this case, u is called the *weak limit* of $\{u_n\}$ and we write $u_n \rightharpoonup u$ in X.

It follows from the Hahn-Banach theorem that the weak limit of a sequence, if it exists, is unique. Also, it is easy to see that the strong convergence implies the weak convergence, i.e., if $u_n \to u$ in X, then $u_n \rightharpoonup u$ in X. The converse of this property is not true in general.

The weak convergence of sequences is used to define weakly closed sets in a normed space.

Definition 1.17. Let X be a normed space. A subset $A \subset X$ is said to be *weakly closed* if it contains the limits of all weakly convergent sequences $\{u_n\} \subset A$.

Clearly, every weakly closed subset of X is closed, but the converse of this property is not true, in general. An important exception is provided by the class of convex sets that is introduced below.

Definition 1.18. Let X be a linear space. A subset $K \subset X$ is said to be *convex* if it has the property

$$u, v \in K \Rightarrow (1-t)\,u + t\,v \in K \quad \forall t \in [0,1].$$

For $t \in [0,1]$, the expression $(1-t)\,u+t\,v$ is said to be a *convex combination* of u and v. The set $\{\,(1-t)\,u+t\,v : t \in [0,1]\,\}$ consists of all the points on the line segment connecting u and v. We see that if K is convex and $u, v \in K$, then the line segment connecting u and v is contained in K.

Theorem 1.19. *A convex subset of a normed space X is closed if and only if it is weakly closed.*

We now introduce the concept of reflexive spaces. To this end, consider a normed space X and denote by $X'' = (X')'$ the dual of the Banach space X', which will be called the *bidual* of X. The bidual X'' is a Banach space. Each element $u \in X$ induces a linear continuous functional $\ell_u \in X''$ by the relation $\ell_u(u') = \langle u', u\rangle$ for every $u' \in X'$. The mapping $u \mapsto \ell_u$ from X into X'' is linear and isometric, i.e., $\|\ell_u\|_{X''} = \|u\|_X$ for all $u \in X$. Therefore, the normed space X may be viewed as a linear subspace of the Banach space X'' by the embedding $u \mapsto \ell_u = \chi(u)$. We introduce the following definition.

Definition 1.20. A normed space X is said to be *reflexive* if X may be identified with X'' by the canonical embedding χ (i.e., if $\chi(X) = X''$).

A reflexive space must be complete and is hence a Banach space. We have the following important property of a reflexive space.

Theorem 1.21. (Eberlein-Smulyan) *If X is a reflexive Banach space, then each bounded sequence in X has a weakly convergent subsequence.*

It follows that if X is a reflexive Banach space and the sequence $\{u_n\} \subset X$ is bounded (i.e., $\sup_n \|u_n\|_X < \infty$), then we can find a subsequence $\{u_{n_k}\} \subset \{u_n\}$ and an element $u \in X$ such that $u_{n_k} \rightharpoonup u$ in X. Furthermore, it can be proved that if the limit u is independent of the subsequence extracted, then the *whole* sequence $\{u_n\}$ converges weakly to u.

On the dual of a normed space, besides the weak convergence, we can introduce the notion of weak * convergence.

Definition 1.22. Let X be a normed space and let X' denote its dual. A sequence $\{u'_n\} \subset X'$ is said to *converge weakly* * to $u' \in X'$ if

$$\langle u'_n, v\rangle \to \langle u', v\rangle \quad \text{as } n \to \infty, \quad \text{for every } v \in X.$$

In this case, u' is called the *weak * limit* of $\{u'_n\}$ and we write $u'_n \rightharpoonup^* u'$ in X'.

It is easy to verify that the weak * limit of a sequence, if it exists, is unique. Moreover, it can be shown that the weak convergence in X' implies the weak * convergence, i.e., if $u'_n \rightharpoonup u'$ in X', then $u'_n \rightharpoonup^* u'$ in X'.

Definition 1.23. A normed space X is said to be *separable* if there exists a countable set $\{v_1, v_2, \dots\} \subset X$ for which the following property is valid: for each $v \in X$, we can find scalars $\{\alpha_{n,i}\}_{i=1}^n$, $n = 1, 2, \dots$, such that

$$\left\| v - \sum_{i=1}^n \alpha_{n,i} v_i \right\|_X \to 0 \quad \text{as } n \to \infty.$$

It follows from the previous definition that finite linear combinations of $\{v_i\}$ are dense in X, as every element in X can be approximated by a sequence of finite linear combinations of $\{v_i\}$. Also, in the definition, every finite number of elements from the set $\{v_i\}$ can be assumed to be linearly independent. We also say that $\{v_1, v_2, \dots\}$ is a countable infinite basis of X.

We have the following important result on separable Banach spaces.

Theorem 1.24. *If X is a separable Banach space, then each bounded sequence in X' has a weakly * convergent subsequence.*

It follows from the previous theorem that if X is a separable Banach space and the sequence $\{u'_n\} \subset X'$ is such that $\sup_n \|u'_n\|_{X'} < \infty$, then we can find a subsequence $\{u'_{n_k}\} \subset \{u'_n\}$ and an element $u' \in X'$ such that $u'_{n_k} \rightharpoonup^* u'$ in X'.

If Y is a subspace of a normed space $(X, \|\cdot\|_X)$, then $(Y, \|\cdot\|_X)$ is a normed space, too. If it is not explicitly stated otherwise, the norm over a subspace is taken to be the norm of the original normed space. Moreover, we have the following theorem.

Theorem 1.25. *Let $(X, \|\cdot\|_X)$ be a Banach space and let $Y \subset X$ be a closed subspace of X. The following results hold:*

(1) Y is a Banach space with the norm $\|\cdot\|_X$.

(2) If X is separable, then Y is separable.

(3) If X is reflexive, then Y is reflexive.

1.3 Hilbert Spaces

In a large number of applications, inner product spaces are usually used. These are the spaces where a norm can be defined through an inner product. The inner product in a general space is a generalization of the usual scalar or dot product in the plane $\mathbb{R}^2$ or the space $\mathbb{R}^3$.

Definition 1.26. Let X be a linear space. An *inner product* $(\cdot,\cdot)_X$ is a function from $X \times X$ to $\mathbb{R}$ with the following properties.

1. $(u,u)_X \geq 0 \quad \forall\, u \in X$, and $(u,u)_X = 0$ iff $u = 0_X$.
2. $(u,v)_X = (v,u)_X \quad \forall\, u,v \in X$.
3. $(\alpha\, u + \beta\, v, w)_X = \alpha\,(u,w)_X + \beta\,(v,w)_X \quad \forall\, u,v,w \in X, \forall\, \alpha,\beta \in \mathbb{R}$.

The pair $(X,(\cdot,\cdot)_X)$ is called an *inner product space*. When the definition of the inner product $(\cdot,\cdot)_X$ is clear from the context, we simply say X is an inner product space.

For an inner product, there is an important property called the *Cauchy-Schwarz inequality*:

$$|(u,v)_X| \leq \sqrt{(u,u)_X\,(v,v)_X} \quad \forall\, u,v \in X,$$

with the equality holding iff u and v are linearly dependent. An inner product $(\cdot,\cdot)_X$ induces a norm through the formula

$$\|u\|_X = \sqrt{(u,u)_X} \quad \forall\, u \in X.$$

Unless stated otherwise, the norm in an inner product space is the one induced by the inner product through the above formula.

Among the inner product spaces, of particular importance are the Hilbert spaces.

Definition 1.27. A complete inner product space is called a *Hilbert space*.

From the definition, we see that an inner product space X is a Hilbert space if X is a Banach space under the norm induced by the inner product.

With the notion of an inner product at our disposal, we can define orthogonality in inner product spaces.

Definition 1.28. We say two vectors u and v are *orthogonal* if $(u,v)_X = 0$. An element u is said to be orthogonal to a subset $A \subset X$ if $(u,v)_X = 0$ for every $v \in A$.

On Hilbert spaces, continuous linear functionals are limited in the forms they can take. The following theorem makes this more precise and represents an important tool in developing solvability theory for variational equations.

Theorem 1.29. (The Riesz Representation Theorem) *Let X be a Hilbert space and let $\ell \in X'$. Then there is a unique $u \in X$ such that*

$$\ell(v) = (u, v)_X \quad \forall\, v \in X.$$

Moreover,

$$\|\ell\|_{X'} = \|u\|_X.$$

By the Riesz representation theorem, we may identify a Hilbert space with its dual. Consequently, a Hilbert space can also be identified with its bidual. Therefore, every Hilbert space is reflexive. Combining this result with Theorem 1.21, we see that each bounded sequence in a Hilbert space has a weakly convergent subsequence.

An important class of nonlinear operators defined in a Hilbert space is provided by the following result.

Theorem 1.30. (The Projection Theorem) *Let K be a nonempty closed convex subset in a Hilbert space X. Then for each $u \in X$, there is a unique element $u_0 = \mathcal{P}_K u \in K$ such that*

$$\|u - u_0\|_X = \min_{v \in K} \|u - v\|_X.$$

The operator $\mathcal{P}_K : X \to K$ is called the *projection operator* onto K. The element $u_0 = \mathcal{P}_K u \in K$ is called the *projection* of u on K and is characterized by the inequality

$$(u_0 - u, v - u_0)_X \geq 0 \quad \forall\, v \in K. \tag{1.3}$$

Using inequality (1.3), it is easy to verify that the projection operator is nonexpansive, that is

$$\|\mathcal{P}_K u - \mathcal{P}_K v\|_X \leq \|u - v\|_X \quad \forall\, u, v \in X,$$

and monotone,

$$(\mathcal{P}_K u - \mathcal{P}_K v, u - v)_X \geq 0 \quad \forall\, u, v \in X.$$

Moreover, if K is a closed subspace of X, then the operator $\mathcal{P}_K : X \to K$ is linear, and the element $u_0 = \mathcal{P}_K u$ is the unique element of K that satisfies

$$(u_0 - u, v)_X = 0 \quad \forall\, v \in K.$$

We conclude that $u - u_0$ is, in this case, orthogonal to K.

Let now $(X, (\cdot,\cdot)_X)$ and $(Y, (\cdot,\cdot)_Y)$ be two Hilbert spaces and denote by $\|\cdot\|_X$ and $\|\cdot\|_Y$ the associated norms. In Section 4.4 of the book, we shall use the product space

$$X \times Y = \{z = (x, y) \,|\, x \in X,\ y \in Y\}$$

together with the canonical inner product

$$(z_1, z_2)_{X\times Y} = (x_1, x_2)_X + (y_1, y_2)_Y$$

for all $z_1 = (x_1, y_1), z_2 = (x_2, y_2) \in X \times Y$. It is easy to verify that $(X \times Y, (\cdot,\cdot)_{X\times Y})$ is a Hilbert space and, moreover,

$$\|z\|^2_{X\times Y} = \|x\|^2_X + \|y\|^2_Y \quad \forall\, z = (x, y) \in X \times Y.$$

1.4 Miscellaneous Results

Convex functions. We start with some results on convex functions defined on inner product spaces. Let $(X, (\cdot,\cdot)_X)$ be such a space and let φ be a function on X with values in $(-\infty, \infty]$, i.e., $\varphi : X \to (-\infty, \infty]$. Below, we adopt the convention that $\infty + \infty = \infty$ and an expression of the form $\infty - \infty$ is undefined. The *effective domain* of φ is the set

$$D(\varphi) = \{u \in X \,:\, \varphi(u) < \infty\},$$

and we say that the function φ is *proper* if $D(\varphi) \neq \emptyset$, that is, there exists $u \in X$ such that $\varphi(u) < \infty$.

Definition 1.31. The function $\varphi : X \to (-\infty, \infty]$ is *convex* if

$$\varphi((1-t)u + tv) \leq (1-t)\varphi(u) + t\varphi(v) \tag{1.4}$$

for all $u, v \in X$ and $t \in [0, 1]$. The function φ is *strictly convex* if the inequality in (1.4) is strict for $u \neq v$ and $t \in (0, 1)$.

We note that if $\varphi, \psi : X \to (-\infty, \infty]$ are convex and $\lambda \geq 0$, then the functions $\varphi + \psi$, $\lambda\varphi$ and *sup* $\{\varphi, \psi\}$ are also convex.

Definition 1.32. The function $\varphi : X \to (-\infty, \infty]$ is said to be *lower semicontinuous* (l.s.c.) at $u \in X$ if

$$\liminf_{n\to\infty} \varphi(u_n) \geq \varphi(u) \tag{1.5}$$

for each sequence $\{u_n\} \subset X$ converging to u in X. The function φ is l.s.c. if it is l.s.c. at every point $u \in X$. When inequality (1.5) holds for each sequence $\{u_n\} \subset X$ that converges weakly to u, the function φ is said to be *weakly lower semicontinuous* at u. The function φ is weakly l.s.c. if it is weakly l.s.c. at every point $u \in X$.

It is easy to see that the norm function $u \mapsto \|u\|_X = (u, u)_X^{\frac{1}{2}}$ is weakly lower semicontinuous. Indeed, let $\{u_n\} \subset X$ be such that $u_n \rightharpoonup u \in X$. For a fixed $v \in X$, the mapping $u \mapsto (v, u)_X$ defines a linear continuous functional on X and therefore

$$\|u\|_X^2 = (u,u)_X = \lim_{n\to\infty} (u,u_n)_X \le \|u\|_X \liminf_{n\to\infty} \|u_n\|_X$$

which implies that

$$\liminf_{n\to\infty} \|u_n\|_X \ge \|u\|_X.$$

If φ is a continuous function then it is also l.s.c. The converse is not true, and a lower semicontinuous function can be discontinuous. Since strong convergence in X implies weak convergence, it follows that a weakly lower semicontinuous function is lower semicontinuous. Moreover, it can be shown that a proper convex function $\varphi : X \to (-\infty, \infty]$ is lower semicontinuous if and only if it is weakly lower semicontinuous.

Let $K \subset X$ and consider the function $\psi_K : X \to (-\infty, \infty]$ defined by

$$\psi_K(v) = \begin{cases} 0 & \text{if } v \in K, \\ \infty & \text{if } v \notin K. \end{cases} \tag{1.6}$$

The function ψ_K is called the *indicator function* of the set K. It can be proved that the set K is a non-empty closed convex set of X if and only if its indicator function ψ_K is a proper convex lower semicontinuous function.

A second example of convex lower semicontinuous function is provided by the following result, which will be used repeatedly in the following chapters of this book.

Proposition 1.33. *Let $(X, (\cdot,\cdot)_X)$ be an inner product space and let $a : X \times X \to \mathbb{R}$ be a bilinear symmetric continuous and X-elliptic form. Then the function $v \mapsto a(v,v)$ is convex and lower semicontinuous.*

Proof. The convexity is straightforward to show. To prove the lower semicontinuity, consider a sequence $\{u_n\} \subset X$ such that $u_n \rightharpoonup u \in X$. Since $a(u_n - u, u_n - u) \ge 0$, it follows that

$$a(u_n,u_n) \ge a(u,u_n) + a(u_n,u) - a(u,u) \quad \forall\, n \in \mathbb{N}. \tag{1.7}$$

For a fixed $v \in X$, the mappings $u \mapsto a(v,u)_X$ and $u \mapsto a(u,v)_X$ define linear continuous functionals on X and therefore

$$\lim_{n\to\infty} a(u,u_n) = \lim_{n\to\infty} a(u_n,u) = a(u,u). \tag{1.8}$$

We pass to the lower limit in (1.7) and use (1.8) to see that

$$\liminf_{n\to\infty} a(u_n,u_n) \ge a(u,u),$$

which concludes the proof. □

By a consequence of the Hahn-Banach theorem, we have the following property of convex lower semicontinuous functions.

Proposition 1.34. *Let $(X, (\cdot,\cdot)_X)$ be an inner product space and let $\varphi : X \to (-\infty, \infty]$ be a proper convex lower semicontinuous function. Then φ is bounded from below by an affine function, i.e., there exists $\alpha \in X$ and $\beta \in \mathbb{R}$ such that $\varphi(u) \geq (\alpha, u)_X + \beta$ for all $u \in X$.*

We now recall the definition of Gâteaux differentiable functions.

Definition 1.35. Let $\varphi : X \to \mathbb{R}$ and let $u \in X$. Then φ is *Gâteaux differentiable* at u if there exists an element $\nabla\varphi(u) \in X$ such that

$$\lim_{t\to 0} \frac{\varphi(u+tv) - \varphi(u)}{t} = (\nabla\varphi(u), v)_X \quad \forall\, v \in X. \tag{1.9}$$

The element $\nabla\varphi(u)$ that satisfies (1.9) is unique and is called the *gradient of φ at u*. The function $\varphi : X \to \mathbb{R}$ is said to be *Gâteaux differentiable* if it is Gâteaux differentiable at every point of X. In this case, the operator $\nabla\varphi : X \to X$ that maps every element $u \in X$ into the element $\nabla\varphi(u)$ is called the *gradient operator* of φ.

The convexity of Gâteaux differentiable functions can be characterized as follows.

Proposition 1.36. *Let $\varphi : X \to \mathbb{R}$ be a Gâteaux differentiable function. Then φ is convex if and only if*

$$\varphi(v) - \varphi(u) \geq (\nabla\varphi(u), v-u)_X \quad \forall\, u,\, v \in X. \tag{1.10}$$

Proof. If φ is a convex function, then for all u, $v \in X$ and $t \in (0,1)$, from (1.4) we deduce that

$$t(\varphi(v) - \varphi(u)) \geq \varphi(u + t(v-u)) - \varphi(u).$$

We divide both sides of the inequality by t, pass to the limit as $t \to 0^+$, and use (1.9) to obtain (1.10). Conversely, assume that (1.10) holds and let u, $v \in X$, $t \in [0,1]$. Let $w = (1-t)u + tv$; it follows from (1.10) that

$$\varphi(v) - \varphi(w) \geq (\nabla\varphi(w), v-w)_X,$$
$$\varphi(u) - \varphi(w) \geq (\nabla\varphi(w), u-w)_X.$$

We multiply the inequalities above by t and $(1-t)$, respectively, then add the results to obtain (1.4), which shows that φ is a convex function. □

From the previous proposition, we easily deduce the following result.

Corollary 1.37. *Let $\varphi : X \to \mathbb{R}$ be a convex Gâteaux differentiable function. Then φ is lower semicontinuous.*

Proof. Let $\{u_n\}$ be a sequence of elements of X converging to u in X. It follows from Proposition 1.36 that

$$\varphi(u_n) - \varphi(u) \geq (\nabla\varphi(u), u_n - u)_X \quad \forall\, n \in \mathbb{N}.$$

We pass to the lower limit as $n \to \infty$ in the previous inequality to obtain (1.5), which concludes the proof. □

Inequality (1.10) suggests a generalization of the gradient operator for convex functions defined on inner product spaces.

Definition 1.38. Let $\varphi : X \to (-\infty, \infty]$ and let $u \in X$. Then the *subdifferential* of φ at u is the set

$$\partial\varphi(u) = \{\, f \in X \,:\, \varphi(v) - \varphi(u) \geq (f, v - u)_X \quad \forall\, v \in X \,\}. \tag{1.11}$$

Denote

$$D(\partial\varphi) = \{u \in X \,:\, \partial\varphi(u) \neq \emptyset\,\}.$$

A function φ is said to be *subdifferentiable* at $u \in X$ if $u \in D(\partial\varphi)$, and each element $f \in \partial\varphi(u)$ is called a *subgradient* of φ at u. A function φ is said to be subdifferentiable if it is subdifferentiable at each point $u \in X$, i.e., if $D(\partial\varphi) = X$.

Using arguments similar to those used in the proof of Corollary 1.37, it can be shown that a subdifferentiable function $\varphi : X \to (-\infty, \infty]$ is convex and lower semicontinuous. Also, for convex functions, the link between the gradient operator and subdifferential is given by the following result.

Proposition 1.39. *Let $\varphi : X \to \mathbb{R}$ be a convex Gâteaux differentiable function. Then φ is subdifferentiable, $\partial\varphi$ is a single-valued operator on X, and $\partial\varphi(u) = \{\nabla\varphi(u)\}$ for every $u \in X$.*

Proof. Let $u \in X$. From Proposition 1.36 and (1.11) we get $\nabla\varphi(u) \in \partial\varphi(u)$, which shows that φ is subdifferentiable at u. Let $f \in \partial\varphi(u)$; using (1.11) we have

$$\varphi(u + tv) - \varphi(u) \geq (f, tv)_X \quad \forall\, v \in X,\ t \in \mathbb{R}.$$

Dividing this inequality by $t > 0$ and using (1.9) we obtain $(\nabla\varphi(u), v)_X \geq (f, v)_X$ for all $v \in X$, which shows that $f = \nabla\varphi(u)$. Therefore, we obtain that $\partial\varphi(u) = \{\nabla\varphi(u)\}$, which concludes the proof. □

The notion of the subdifferential is useful in describing constraints in many branches of engineering and in mechanics, including those arising in contact problems, as it will be shown in Section 7.5.

Fixed point theorems. We continue with fixed point results that will be used later in this book.

Theorem 1.40. (The Banach Fixed Point Theorem) *Let K be a nonempty closed subset of a Banach space $(X, \|\cdot\|_X)$. Assume that $\Lambda : K \to K$ is a* contraction, *i.e., there exists a constant $\alpha \in [0,1)$ such that*

$$\|\Lambda u - \Lambda v\|_X \le \alpha \|u - v\|_X \quad \forall\, u, v \in K.$$

Then there exists a unique $u \in K$ such that $\Lambda u = u$.

A solution $u \in K$ of the operator equation $\Lambda u = u$ is called a *fixed point* of Λ in K. We also need a variant of the Banach fixed point theorem, which we recall next. To this end, for an operator Λ, we define its powers inductively by the formula $\Lambda^m = \Lambda(\Lambda^{m-1})$ for $m \ge 2$.

Theorem 1.41. *Assume that K is a nonempty closed set in a Banach space X and let $\Lambda : K \to K$. Assume also that $\Lambda^m : K \to K$ is a contraction for some positive integer m. Then Λ has a unique fixed-point in K.*

We now provide a result that will be used repeatedly in Part II of the book to prove existence results for the solutions of variational inequalities.

Let $(X, \|\cdot\|_X)$ be a Banach space, $T > 0$, and denote by $C([0,T];X)$ the space of continuous functions defined on $[0,T]$ with values in X. It is well known that $C([0,T];X)$ is a Banach space with the norm

$$\|v\|_{C([0,T];X)} = \max_{t\in[0,T]} \|v(t)\|_X.$$

Lemma 1.42. *Let $\Lambda : C([0,T];X) \to C([0,T];X)$ be an operator that satisfies the following property: there exists $c > 0$ such that*

$$\|\Lambda\eta_1(t) - \Lambda\eta_2(t)\|_X \le c \int_0^t \|\eta_1(s) - \eta_2(s)\|_X\, ds \qquad \forall\, \eta_1, \eta_2 \in C([0,T];X), \quad t \in [0,T]. \tag{1.12}$$

Then, there exists a unique element $\eta^* \in C([0,T];X)$ such that $\Lambda\eta^* = \eta^*$.

Proof. A first proof of the lemma can be obtained as follows. Let $\eta_1, \eta_2 \in C([0,T];X)$ and let $t \in [0,T]$. Reiterating (1.12), it follows that

$$\|\Lambda^m\eta_1(t) - \Lambda^m\eta_2(t)\|_X \le c^m \underbrace{\int_0^t \int_0^s \ldots \int_0^u}_{m \text{ integrals}} \|\eta_1(r) - \eta_2(r)\|_X\, dr \ldots ds,$$

which implies that

$$\|\Lambda^m\eta_1 - \Lambda^m\eta_2\|_{C([0,T];X)} \le \frac{c^m T^m}{m!} \|\eta_1 - \eta_2\|_{C([0,T];X)}. \tag{1.13}$$

Since

$$\lim_{m\to\infty} \frac{c^m T^m}{m!} = 0,$$

it follows that there exists a positive integer m such that $\frac{c^m T^m}{m!} < 1$ and, therefore, (1.13) shows that Λ^m is a contraction on the Banach space $C([0,T];X)$. Lemma 1.42 is now a consequence of Theorem 1.41.

A second proof of the lemma, which avoids the use of Theorem 1.41, is the following: denote

$$\|\eta\|_\beta = \max_{t\in[0,T]} e^{-\beta t}\, \|\eta(t)\|_X \quad \forall\, \eta \in C([0,T];X),$$

with $\beta > 0$ to be chosen later. Clearly, $\|\cdot\|_\beta$ defines a norm on the space $C([0,T];X)$ that is equivalent to the standard norm $\|\cdot\|_{C([0,T];X)}$. Using (1.12), after some calculus we find that

$$\|\Lambda\eta_1 - \Lambda\eta\|_\beta \le \frac{c}{\beta}\, \|\eta_1 - \eta_2\|_\beta \quad \forall\, \eta_1,\, \eta_2 \in C([0,T];X).$$

So, if we choose β such that $\beta > c$, then the operator Λ is a contraction on the space $C([0,T];X)$ endowed with the equivalent norm $\|\cdot\|_\beta$. By the Banach fixed point theorem (Theorem 1.40), it follows that Λ has a unique fixed point $\eta^* \in C([0,T];X)$, which concludes the proof. □

We continue with the following fixed point result.

Theorem 1.43. (Schauder-Tychonoff) *Let X be a reflexive separable Banach space and let K be a nonempty closed bounded convex subset of X. Let also $\Lambda : K \to K$ be a weakly sequentially continuous map, that is, for each sequence $\{x_n\} \subset K$, $x_n \rightharpoonup x$ in $X \Longrightarrow \Lambda x_n \rightharpoonup \Lambda x$ in X. Then, there exists an element $x \in K$ such that $\Lambda x = x$.*

Some elementary inequalities. We introduce now some elementary inequalities that will be used in later chapters. To this end, we use below the notation $C([a,b])$ for the space of real valued continuous functions defined on the interval $[a,b] \subset \mathbb{R}$.

The following Gronwall inequality will be used frequently in the study of evolutionary variational inequalities.

Lemma 1.44. (Gronwall's Inequality) *Assume $f, g \in C([a,b])$ satisfying*

$$f(t) \le g(t) + c\int_a^t f(s)\, ds \quad \forall\, t \in [a,b], \tag{1.14}$$

where $c > 0$ is a constant. Then

$$f(t) \le g(t) + c\int_a^t g(s)\, e^{c\,(t-s)}\, ds \quad \forall\, t \in [a,b]. \tag{1.15}$$

Moreover, if g is nondecreasing, then

$$f(t) \leq g(t)\, e^{c\,(t-a)} \quad \forall\, t \in [a,b]. \tag{1.16}$$

Proof. Define

$$F(s) = \int_a^s f(r)\, dr \quad \forall\, s \in [a,b]. \tag{1.17}$$

Then $F'(s) = f(s)$. From assumption (1.14) we get

$$F'(s) \leq g(s) + c\, F(s) \quad \forall\, s \in [a,b].$$

Thus,

$$\left(e^{-c\,s} F(s)\right)' \leq g(s)\, e^{-c\,s} \quad \forall\, s \in [a,b].$$

Let $t \in [a,b]$. We integrate this inequality from a to t and note that $F(a) = 0$; as a result we find

$$e^{-c\,t} F(t) \leq \int_a^t g(s)\, e^{-c\,s} ds.$$

So,

$$F(t) \leq \int_a^t g(s)\, e^{c\,(t-s)} ds. \tag{1.18}$$

We combine now (1.14) with (1.17) to find that $f(t) \leq g(t) + cF(t)$, then, using (1.18), we obtain (1.15).

If g is nondecreasing, then (1.15) implies that

$$f(t) \leq g(t) + g(t)\, c \int_a^t e^{c\,(t-s)}\, ds = g(t)\, e^{c\,(t-a)},$$

which concludes the proof. □

We also need the following result.

Lemma 1.45. (Young's Inequality) *Let p, $q \in \mathbb{R}$ be two conjugate exponents, that is $1 < p < \infty$ and $\frac{1}{p} + \frac{1}{q} = 1$. Then*

$$ab \leq \frac{a^p}{p} + \frac{b^q}{q} \quad \forall\, a,\, b \geq 0. \tag{1.19}$$

Proof. Let a, $b \geq 0$. Since (1.19) is obvious in the case $a = 0$ or $b = 0$, we assume in what follows that $a > 0$ and $b > 0$. Consider the function

$f : (0, \infty) \to \mathbb{R}$ given by

$$f(t) = t^{\frac{1}{p}} - \frac{1}{p}\, t.$$

We have

$$f'(t) = \frac{1}{p}\big(t^{\frac{1}{p}-1} - 1\big), \quad f''(t) = \frac{1}{p}\Big(\frac{1}{p} - 1\Big) t^{\frac{1}{p}-2}$$

and, since $p > 1$, it follows that the second derivative f'' is negative on $(0, \infty)$ and therefore the first derivative f' is a decreasing function. Equality $f'(1) = 0$ implies that f' is positive on $(0, 1)$ and negative on $(1, \infty)$, which shows that f has a global maximum at $t = 1$. We conclude that $f(t) \leq f(1)$ for all $t \geq 0$, which implies that

$$t^{\frac{1}{p}} - 1 \leq \frac{1}{p}\,(t-1) \quad \forall\, t > 0. \tag{1.20}$$

We take $t = \frac{a^p}{b^q}$ in (1.20) and, after some elementary manipulation, using the identity $\frac{1}{p} + \frac{1}{q} = 1$ we obtain (1.19). □

Lemma 1.45 allows us to obtain the following result.

Lemma 1.46. *Let $p > 0$. Then*

$$|u|^{p-1}u(v-u) \leq \frac{1}{p+1}\,|v|^{p+1} - \frac{1}{p+1}\,|u|^{p+1} \quad \forall\, u,\, v \in \mathbb{R},\ u \neq 0. \tag{1.21}$$

Proof. Let $u,\, v \in \mathbb{R}$, $u \neq 0$. It is easy to see that

$$|u|^{p-1}u(v-u) \leq |u|^p|v| - |u|^{p+1}. \tag{1.22}$$

Next, applying (1.19) with $a = |v|$, $b = |u|^p$ and with the pair of conjugate exponents $(p+1, \frac{p+1}{p})$, we have

$$|u|^p|v| \leq \frac{1}{p+1}\,|v|^{p+1} + \frac{p}{p+1}\,|u|^{p+1}. \tag{1.23}$$

Inequality (1.21) is now a consequence of (1.22) and (1.23). □

Chapter 2

Function Spaces

We introduce in this chapter the function spaces that will be relevant to the subsequent developments in this monograph. The function spaces to be discussed include spaces of continuous and continuously differentiable functions, Lebegue and Sobolev spaces, associated with an open bounded domain $\Omega \subset \mathbb{R}^d$. In order to treat time-dependent problems, we also introduce spaces of vector-valued functions, i.e., spaces of mappings defined on a time interval $[0, T] \subset \mathbb{R}$ with values into a Banach or Hilbert space X.

2.1 The Spaces $C^m(\overline{\Omega})$ and $L^p(\Omega)$

Let Ω be an open bounded subset of $\mathbb{R}^d$, where d is a positive integer. We denote by Γ the boundary of Ω, and $\overline{\Omega} = \Omega \cup \Gamma$ the closure of Ω. A typical point in $\mathbb{R}^d$ is denoted by $\boldsymbol{x} = (x_1, \ldots, x_d)$ or $\boldsymbol{x} = (x_1, \ldots, x_d)^T$. It is convenient to use the multi-index notation for partial derivatives. An ordered collection of d non-negative integers, $\alpha = (\alpha_1, \ldots, \alpha_d)$, is called a *multi-index*. The quantity $|\alpha| = \sum_{i=1}^d \alpha_i$ is said to be the *length* of α. If v is an m-times differentiable function, then for each α with $|\alpha| \le m$,

$$D^\alpha v(\boldsymbol{x}) = \frac{\partial^{|\alpha|} v(\boldsymbol{x})}{\partial x_1^{\alpha_1} \cdots \partial x_d^{\alpha_d}}$$

is the αth partial derivative. This is a handy notation for partial derivatives. Some examples are

$$\partial_1 v = \frac{\partial v}{\partial x_1} = D^\alpha v, \quad \alpha = (1, 0, \ldots, 0),$$
$$\frac{\partial^d v}{\partial x_1 \cdots \partial x_d} = D^\alpha v, \quad \alpha = (1, 1, \ldots, 1).$$

The set of all the partial derivatives of order m of a function v can be expressed as $\{ D^\alpha v : |\alpha| = m \}$. There are other notations commonly used for partial derivatives; e.g., the partial derivative $\partial v/\partial x_i$ is also written as $\partial_{x_i} v$, or $\partial_i v$, or v_{x_i}, or $v_{,i}$.

The spaces $C^m(\overline{\Omega})$. Let $C(\overline{\Omega})$ be the space of functions that are uniformly continuous on Ω. Each function in $C(\overline{\Omega})$ is bounded. The notation $C(\overline{\Omega})$ is consistent with the fact that a uniformly continuous function on Ω has a unique continuous extension to $\overline{\Omega}$. For $v \in C(\overline{\Omega})$, we always understand its boundary values on Γ to be the continuous extension of the values of v in Ω. The space $C(\overline{\Omega})$ is a Banach space with the norm

$$\|v\|_{C(\overline{\Omega})} = \sup\{ |v(\boldsymbol{x})| : \boldsymbol{x} \in \Omega \} \equiv \max\{ |v(\boldsymbol{x})| : \boldsymbol{x} \in \overline{\Omega} \}.$$

Denote $\mathbb{Z}_+$ the set of non-negative integers. For every $m \in \mathbb{Z}_+$, $C^m(\overline{\Omega})$ is the space of functions that, together with their derivatives of order less than or equal to m, are continuous up to the boundary,

$$C^m(\overline{\Omega}) = \{ v \in C(\overline{\Omega}) : D^\alpha v \in C(\overline{\Omega}) \text{ for } |\alpha| \le m \}.$$

When $m = 0$, we usually write $C(\overline{\Omega})$ instead of $C^0(\overline{\Omega})$. The space $C^m(\overline{\Omega})$ is a Banach space with the norm

$$\|v\|_{C^m(\overline{\Omega})} = \sum_{|\alpha| \le m} \|D^\alpha v\|_{C(\overline{\Omega})}.$$

We also set

$$C^\infty(\overline{\Omega}) = \bigcap_{m=0}^{\infty} C^m(\overline{\Omega}) \equiv \{ v \in C(\overline{\Omega}) : v \in C^m(\overline{\Omega}) \quad \forall\, m \in \mathbb{Z}_+ \},$$

the space of infinitely differentiable functions.

Given a function $v : \Omega \to \mathbb{R}$, its *support* is defined to be

$$\operatorname{supp} v = \overline{\{ \boldsymbol{x} \in \Omega : v(\boldsymbol{x}) \ne 0 \}}.$$

We say that v has a *compact support* if $\operatorname{supp} v$ is a proper subset of Ω: $\operatorname{supp} v \subset \Omega$. Thus, if v has a compact support, then there is a neighboring open strip about the boundary Γ such that v is zero on the part of the strip that lies inside Ω. Later on, we will need the space

$$C_0^\infty(\Omega) = \{ v \in C^\infty(\Omega) : \operatorname{supp} v \subset \Omega \}.$$

Obviously, $C_0^\infty(\Omega) \subset C^\infty(\overline{\Omega})$.

The spaces $L^p(\Omega)$. For every number $p \in [1, \infty)$ we denote by $L^p(\Omega)$ the linear space of (equivalence classes of) measurable functions $v : \Omega \to \mathbb{R}$

for which

$$\|v\|_{L^p(\Omega)} = \left(\int_\Omega |v(\boldsymbol{x})|^p dx \right)^{1/p} < \infty, \tag{2.1}$$

where the integration is understood to be in the sense of Lebesgue. The mapping $v \mapsto \|v\|_{L^p(\Omega)}$ is a norm on $L^p(\Omega)$. Here and below, it is understood that v represents an equivalence class of functions, two functions being equivalent if they are equal almost everywhere (a.e.), that is, equal except on a subset of Ω of Lebesgue measure zero.

The definition of the spaces $L^p(\Omega)$ can be extended to include the case $p = \infty$ in the following manner. For every measurable function $v : \Omega \to \mathbb{R}$ denote

$$\begin{aligned} \|v\|_{L^\infty(\Omega)} &= \operatorname*{ess\,sup}_{\boldsymbol{x}\in\Omega} |v(\boldsymbol{x})| \\ &= \inf\{\, M \in [0,\infty] \,:\, |v(\boldsymbol{x})| \le M \quad \text{a.e. } \boldsymbol{x} \in \Omega \,\}. \end{aligned} \tag{2.2}$$

The quantity (2.2) is called the *essential supremum* of $|v|$, and we say that v is *essentially bounded* if $\|v\|_{L^\infty(\Omega)} < \infty$. We denote by $L^\infty(\Omega)$ the linear space of (equivalence classes of) measurable functions $v : \Omega \to \mathbb{R}$ that are essentially bounded. The mapping $v \mapsto \|v\|_{L^\infty(\Omega)}$ is a norm on the space $L^\infty(\Omega)$.

For $p \in [1,\infty]$, its conjugate exponent $q \in [1,\infty]$ is defined by the relations

$$\begin{aligned} \frac{1}{p} + \frac{1}{q} &= 1 \quad \text{if } p \neq 1, \\ q &= \infty \quad \text{if } p = 1. \end{aligned}$$

Here, we adopt the convention $1/\infty = 0$ and therefore $q = 1$ if $p = \infty$.

Some basic properties of the L^p spaces are summarized below.

Theorem 2.1. *Let Ω be an open bounded set in $\mathbb{R}^d$ and let $p \in [1,\infty]$. Then:*

(a) *$L^p(\Omega)$ is a Banach space.*
(b) *Every Cauchy sequence in $L^p(\Omega)$ has a subsequence that converges pointwise a.e. on Ω.*
(c) (Hölder's inequality) *Let $u \in L^p(\Omega)$, $v \in L^q(\Omega)$. Then*

$$\int_\Omega |u(\boldsymbol{x})\, v(\boldsymbol{x})|\, dx \le \|u\|_{L^p(\Omega)} \|v\|_{L^q(\Omega)}.$$

(d) *For $1 \le p < \infty$, the dual $(L^p(\Omega))'$ of $L^p(\Omega)$ is the space $L^q(\Omega)$. Moreover, if $p \in (1,\infty)$, the space $L^p(\Omega)$ is reflexive.*
(e) *For $1 \le p < \infty$, $L^p(\Omega)$ is a separable space.*

The case $p = 2$ is special since in this case $p = q = 2$. A simple consequence of the last theorem is the following result.

Corollary 2.2. *The space $L^2(\Omega)$ is a Hilbert space with the inner product*

$$(u, v)_{L^2(\Omega)} = \int_\Omega u(\boldsymbol{x})\, v(\boldsymbol{x})\, dx \quad \forall\, u, v \in L^2(\Omega).$$

Moreover, $L^2(\Omega)$ is separable and the following Cauchy-Schwarz inequality holds:

$$\left| \int_\Omega u(\boldsymbol{x})\, v(\boldsymbol{x})\, dx \right| \le \|u\|_{L^2(\Omega)} \|v\|_{L^2(\Omega)} \quad \forall\, u, v \in L^2(\Omega).$$

We now introduce the spaces of locally integrable functions that will be used in the next section in order to introduce the weak derivatives of real-valued functions defined on Ω.

Definition 2.3. Let $1 \le p < \infty$. A function $v : \Omega \subset \mathbb{R}^d \to \mathbb{R}$ is said to be *locally p-integrable*, $v \in L^p_{\mathrm{loc}}(\Omega)$, if for every $\boldsymbol{x} \in \Omega$, there is an open neighborhood Ω' of $\boldsymbol{x}$ such that $\Omega' \subset\subset \Omega$ (i.e., $\overline{\Omega'} \subset \Omega$) and $v \in L^p(\Omega')$.

We have the following useful result ([153, p. 18]).

Lemma 2.4. (Generalized Variational Lemma) *Let $v \in L^1_{\mathrm{loc}}(\Omega)$ with Ω a nonempty open set in $\mathbb{R}^d$. If*

$$\int_\Omega v(\boldsymbol{x})\, \phi(\boldsymbol{x})\, dx = 0 \quad \forall\, \phi \in C_0^\infty(\Omega),$$

then $v = 0$ a.e. on Ω.

In Section 2.3, we shall need the space

$$L^2(\Omega)^d = \{\, \boldsymbol{u} = (u_i) \,:\, u_i \in L^2(\Omega),\, i = 1, \ldots, d \,\}. \tag{2.3}$$

This is a Hilbert space with the canonical inner product

$$(\boldsymbol{u}, \boldsymbol{v})_{L^2(\Omega)^d} = \sum_{i=1}^d \int_\Omega u_i(\boldsymbol{x}) v_i(\boldsymbol{x})\, dx$$

and the associated norm

$$\|\boldsymbol{u}\|_{L^2(\Omega)^d} = \left(\sum_{i=1}^d \int_\Omega u_i^2(\boldsymbol{x})\, dx \right)^{\frac{1}{2}}.$$

2.2 Sobolev Spaces

Sobolev spaces are indispensable tools in the study of boundary value problems. To introduce Sobolev spaces, we first need to extend the definition of derivatives. The starting point is the classical "integration by parts" formula

$$\int_{\Omega} v(\boldsymbol{x})\, D^{\alpha}\phi(\boldsymbol{x})\, dx = (-1)^{|\alpha|} \int_{\Omega} D^{\alpha}v(\boldsymbol{x})\, \phi(\boldsymbol{x})\, dx, \tag{2.4}$$

which holds for $v \in C^k(\Omega)$, $\phi \in C_0^{\infty}(\Omega)$ and $|\alpha| \leq k$. This formula, relating differentiation and integration, is the most important formula in calculus. The *weak derivative* is an extension of the classical derivative, i.e., if the classical derivative exists, then the two derivatives coincide. To ensure that the weak derivative is useful, we require that the integration by parts formula (2.4) holds. A more general approach for the extension of the classical derivatives is to first introduce the derivatives in the distributional sense. A detailed discussion of distributions and the derivatives in the distributional sense can be found in several monographs, e.g., [138]. Here we choose to introduce the concept of the weak derivatives directly, which is sufficient for this work.

Definition 2.5. Let Ω be a nonempty open set in $\mathbb{R}^d$, $v, w \in L^1_{\text{loc}}(\Omega)$. Then w is called an αth *weak derivative* of v if

$$\int_{\Omega} v(\boldsymbol{x})\, D^{\alpha}\phi(\boldsymbol{x})\, dx = (-1)^{|\alpha|} \int_{\Omega} w(\boldsymbol{x})\, \phi(\boldsymbol{x})\, dx \quad \forall\, \phi \in C_0^{\infty}(\Omega). \tag{2.5}$$

Applying Lemma 2.4, we see that when a weak derivative exists, it is uniquely defined up to a set of measure zero. Also, from the definition of the weak derivative, we see that if v is k-times continuously differentiable on Ω, then, for each α with $|\alpha| \leq k$, the classical partial derivative $D^{\alpha}v$ is also the αth weak derivative of v. For this reason, we use the notation $D^{\alpha}v$ also for the αth weak derivative of v and we note that $D^{\alpha}v = v$ if $|\alpha| = 0$.

We continue with the following definition.

Definition 2.6. Let k be a non-negative integer, $k \in \mathbb{Z}_+$, and let $p \in [1, \infty]$. The *Sobolev space* $W^{k,p}(\Omega)$ is the set of all the functions $v \in L^1_{\text{loc}}(\Omega)$ such that for each multi-index α with $|\alpha| \leq k$, the αth weak derivative $D^{\alpha}v$ exists and belongs to $L^p(\Omega)$. The norm in the space $W^{k,p}(\Omega)$ is defined as

$$\|v\|_{W^{k,p}(\Omega)} = \begin{cases} \left(\displaystyle\sum_{|\alpha| \leq k} \|D^{\alpha}v\|^p_{L^p(\Omega)} \right)^{1/p} & \text{if } \ 1 \leq p < \infty, \\ \displaystyle\max_{|\alpha| \leq k} \|D^{\alpha}v\|_{L^{\infty}(\Omega)} & \text{if } \ p = \infty. \end{cases}$$

We note that $W^{0,p}(\Omega) = L^p(\Omega)$, and when $p = 2$, we write $W^{k,2}(\Omega) \equiv H^k(\Omega)$.

A seminorm over the space $W^{k,p}(\Omega)$ is

$$|v|_{W^{k,p}(\Omega)} = \begin{cases} \left(\displaystyle\sum_{|\alpha| = k} \|D^{\alpha}v\|^p_{L^p(\Omega)} \right)^{1/p} & \text{if } \ 1 \leq p < \infty, \\ \displaystyle\max_{|\alpha| = k} \|D^{\alpha}v\|_{L^{\infty}(\Omega)} & \text{if } \ p = \infty. \end{cases}$$

It is not difficult to see that $W^{k,p}(\Omega)$ is a normed space. Moreover, we have the following results, which summarize the basic properties of Sobolev spaces.

Theorem 2.7. *Let Ω be an open bounded set in $\mathbb{R}^d$, $k \in \mathbb{Z}_+$ and $p \in [1, \infty]$. Then:*

(a) *$W^{k,p}(\Omega)$ is a Banach space.*
(b) *$W^{k,p}(\Omega)$ is reflexive if $1 < p < \infty$.*
(c) *$W^{k,p}(\Omega)$ is separable if $1 \leq p < \infty$.*

A simple consequence of the Theorem 2.7 combined with Corollary 2.2 is the following result.

Corollary 2.8. *The Sobolev space $H^k(\Omega)$ is a separable Hilbert space with the inner product*

$$(u, v)_{H^k(\Omega)} = \int_\Omega \sum_{|\alpha| \leq k} D^\alpha u(\boldsymbol{x})\, D^\alpha v(\boldsymbol{x})\, dx \quad \forall\, u, v \in H^k(\Omega).$$

The closure of the space $C_0^\infty(\Omega)$ with respect to the norm $\|\cdot\|_{W^{k,p}(\Omega)}$ gives a closed subspace of $W^{k,p}(\Omega)$, denoted $W_0^{k,p}(\Omega)$. When $p = 2$, we use the notation $H_0^k(\Omega) \equiv W_0^{k,2}(\Omega)$. It follows from Theorem 2.7 and Corollary 2.8 that $W_0^{k,p}(\Omega)$ is a Banach space and $H_0^k(\Omega)$ is a Hilbert space. It can be shown that the seminorm $|\cdot|_{W^{k,p}(\Omega)}$ is a norm on $W_0^{k,p}(\Omega)$ and there exists a constant $c > 0$ such that

$$|v|_{W^{k,p}(\Omega)} \leq \|v\|_{W^{k,p}(\Omega)} \leq c\, |v|_{W^{k,p}(\Omega)} \quad \forall\, v \in W_0^{k,p}(\Omega).$$

We now collect some important properties of Sobolev spaces. Some of them require a certain degree of regularity on the boundary Γ of the domain Ω and, for this reason, we introduce the following definition.

Definition 2.9. Let Ω be open and bounded in $\mathbb{R}^d$. We say that Ω has a *Lipschitz continuous boundary* Γ if for each point $\boldsymbol{x}_0 \in \Gamma$ there exists $r > 0$ and a Lipschitz continuous function $g : \mathbb{R}^{d-1} \to \mathbb{R}$ such that, upon relabeling the coordinate axes if necessary, we have

$$\Omega \cap B(\boldsymbol{x}_0, r) = \{\, \boldsymbol{x} \in B(\boldsymbol{x}_0, r) : x_d > g(x_1, \ldots, x_{d-1}) \,\}.$$

Here, $B(\boldsymbol{x}_0, r)$ denotes the ball of $\mathbb{R}^d$ centered at $\boldsymbol{x}_0$ with radius r. Also, we recall that a real-valued function $g : \mathbb{R}^{d-1} \to \mathbb{R}$ is said to be *Lipschitz continuous* if for some constant L_g, there holds the inequality

$$|g(\boldsymbol{x}) - g(\boldsymbol{y})| \leq L_g \,\|\boldsymbol{x} - \boldsymbol{y}\| \quad \forall\, \boldsymbol{x}, \boldsymbol{y} \in \mathbb{R}^{d-1},$$

where $\|\boldsymbol{x}-\boldsymbol{y}\|$ denotes the standard Euclidean distance between $\boldsymbol{x}$ and $\boldsymbol{y}$,

$$\|\boldsymbol{x}-\boldsymbol{y}\| = \left(\sum_{i=1}^{d-1} |x_i - y_i|^2\right)^{1/2}.$$

We note that, since Γ is a compact set in $\mathbb{R}^d$, in Definition 2.9 we can actually find a finite number of points $\{\boldsymbol{x}_i\}_{i=1}^N$ on the boundary, positive numbers $\{r_i\}_{i=1}^N$ and Lipschitz continuous functions $\{g_i\}_{i=1}^N$, such that Γ is covered by the union of the balls $B(\boldsymbol{x}_i, r_i)$, $1 \le i \le N$, and

$$\Omega \cap B(\boldsymbol{x}_i, r_i) = \{\, \boldsymbol{x} \in B(\boldsymbol{x}_i, r_i) : x_d > g_i(x_1, \ldots, x_{d-1})\,\}$$

upon relabeling the coordinate axes, if necessary.

With a slight abuse of terminology, we say that a domain Ω is a *Lipschitz domain* if it is an open bounded domain with Lipschitz continuous boundary. In the following, we always assume that Ω is a Lipschitz domain, and we note that this assumption is, in fact, not needed for some of the results stated below.

We have the following result on the approximation of Sobolev functions by smooth functions.

Theorem 2.10. *Assume that $\Omega \subset \mathbb{R}^d$ is a Lipschitz domain, $k \in \mathbb{Z}_+$ and $p \in [1,\infty)$. Then for each $v \in W^{k,p}(\Omega)$, there exists a sequence $\{v_n\} \subset C^\infty(\overline{\Omega})$ such that*

$$\|v_n - v\|_{W^{k,p}(\Omega)} \to 0 \quad \text{as } n \to \infty.$$

From the definition of the space $W_0^{k,p}(\Omega)$, we immediately obtain the following density result.

Theorem 2.11. *Under the assumptions of Theorem 2.10, for every $v \in W_0^{k,p}(\Omega)$, there exists a sequence $\{v_n\} \subset C_0^\infty(\Omega)$ such that*

$$\|v_n - v\|_{W^{k,p}(\Omega)} \to 0 \quad \text{as } n \to \infty.$$

To compare Sobolev spaces with different indices, we need the following definitions.

Definition 2.12. Let X and Y be two normed spaces with $X \subset Y$. We say the space X is *continuously embedded* in Y and write $X \hookrightarrow Y$, if the identity operator $I : X \to Y$ is continuous. We say the space X is *compactly embedded* in Y and write $X \hookrightarrow\hookrightarrow Y$, if the identity operator $I : X \to Y$ is compact.

It is easy to see that $X \hookrightarrow Y$ if and only if there exists $c > 0$ such that

$$\|v\|_Y \le c\,\|v\|_X \quad \forall\, v \in X \tag{2.6}$$

or, equivalently,

$$\forall \{v_n\} \subset X, \quad v_n \to v \text{ in } X \Longrightarrow v_n \to v \text{ in } Y.$$

Also, it can be proved that $X \hookrightarrow\hookrightarrow Y$ iff

$$\forall \{v_n\} \subset X, \quad v_n \rightharpoonup v \text{ in } X \Longrightarrow v_n \to v \text{ in } Y.$$

Some properties regarding embeddings and compact embeddings involving Sobolev spaces, which are important in analyzing the regularity of weak solutions of boundary value problems, are summarized in the following theorem.

Theorem 2.13. *Let $\Omega \subset \mathbb{R}^d$ be a Lipschitz domain, $k \in \mathbb{Z}_+$ and $p \in [1, \infty)$. Then the following statements are valid.*
(a) *If $k < \frac{d}{p}$, then $W^{k,p}(\Omega) \hookrightarrow L^q(\Omega)$ for every $q \leq p^*$ and $W^{k,p}(\Omega) \hookrightarrow\hookrightarrow L^q(\Omega)$ for every $q < p^*$, where $\frac{1}{p^*} = \frac{1}{p} - \frac{k}{d}$.*
(b) *If $k = \frac{d}{p}$, then $W^{k,p}(\Omega) \hookrightarrow\hookrightarrow L^q(\Omega)$ for every $q < \infty$.*
(c) *If $k > \frac{d}{p}$, then $W^{k,p}(\Omega) \hookrightarrow\hookrightarrow C^m(\overline{\Omega})$ for every integer m that satisfies $0 \leq m < k - \frac{d}{p}$.*

A direct consequence of Theorem 2.13 is the following result.

Corollary 2.14. *Let $\Omega \subset \mathbb{R}^d$ be a Lipschitz domain, $p \in [1, \infty)$ and let k be a positive integer. Then $W^{k,p}(\Omega) \hookrightarrow\hookrightarrow L^p(\Omega)$.*

In this work, we shall use the embedding result

$$H^1(\Omega) \hookrightarrow\hookrightarrow L^2(\Omega) \tag{2.7}$$

which is obtained from Corollary 2.14 by taking $k = 1$ and $p = 2$.

Sobolev spaces are defined through $L^p(\Omega)$ spaces. Hence Sobolev functions are uniquely defined only a.e. in Ω. Since the boundary Γ has measure zero in $\mathbb{R}^d$, the boundary values of a Sobolev function seem to be not well defined. Nevertheless, it is possible to define the trace of a Sobolev function on the boundary in such a way that for a Sobolev function that is continuous up to the boundary, its trace coincides with its boundary value.

Theorem 2.15. *Assume that Ω is a Lipschitz domain in $\mathbb{R}^d$ with boundary Γ and $1 \leq p < \infty$. Then there exists a linear continuous operator $\gamma : W^{1,p}(\Omega) \to L^p(\Gamma)$ such that $\gamma v = v|_\Gamma$ if $v \in W^{1,p}(\Omega) \cap C(\overline{\Omega})$. Moreover, the mapping $\gamma : W^{1,p}(\Omega) \to L^p(\Gamma)$ is compact, i.e., for every bounded sequence $\{v_n\}$ in $W^{1,p}(\Omega)$, there is a subsequence $\{v_{n_k}\} \subset \{v_n\}$ such that $\{\gamma v_{n_k}\}$ is convergent in $L^p(\Gamma)$.*

The operator γ is called the *trace operator*, and γv can be termed the *trace* of $v \in W^{1,p}(\Omega)$. For the sake of simplicity, when no ambiguity occurs, we usually write v instead of γv. Note that the continuity of γ implies that there exists a constant $c > 0$, which depends on Ω, such that

$$\|\gamma v\|_{L^p(\Gamma)} \leq c\,\|v\|_{W^{1,p}(\Omega)} \quad \forall\, v \in W^{1,p}(\Omega). \tag{2.8}$$

Generally, the trace operator is neither an injection nor a surjection from $W^{1,p}(\Omega)$ to $L^p(\Gamma)$ (the only exception is when $p = 1$, since then the trace operator is surjective from $W^{1,1}(\Omega)$ to $L^1(\Gamma)$). When $p > 1$, the range $\gamma(W^{1,p}(\Omega))$ is a space smaller than $L^p(\Gamma)$.

2.3 Equivalent Norms on the Space $H^1(\Omega)$

In studying second-order boundary value problems, the Sobolev space $H^1(\Omega)$ plays a crucial role. For this reason, since the antiplane contact problems we study in Part IV of this monograph lead to second-order boundary value problems, in this section we pay a particular attention to the Sobolev space $H^1(\Omega)$.

Let Ω be a Lipschitz domain in $\mathbb{R}^d$. Recall that

$$H^1(\Omega) = \{u \in L^2(\Omega) \,:\, \nabla u \in L^2(\Omega)^d\,\}$$

where $\nabla u = (\partial_1 u, \ldots, \partial_d u)$ represents the *gradient operator*. The space $H^1(\Omega)$ is a Hilbert space with the inner product

$$(u, v)_{H^1(\Omega)} = (u, v)_{L^2(\Omega)} + (\nabla u, \nabla v)_{L^2(\Omega)^d} \tag{2.9}$$

and the associated norm

$$\|u\|_{H^1(\Omega)} = \big(\|u\|^2_{L^2(\Omega)} + \|\nabla u\|^2_{L^2(\Omega)^d}\big)^{\frac{1}{2}}. \tag{2.10}$$

A seminorm over the space $H^1(\Omega)$ is

$$|u|_{H^1(\Omega)} = \|\nabla u\|_{L^2(\Omega)^d}.$$

It also follows from Theorem 2.15 that the trace map $\gamma : H^1(\Omega) \to L^2(\Gamma)$ is a linear continuous and compact operator. In what follows, we simply write v for the trace γv of an element $v \in H^1(\Omega)$ and we note that inequality (2.8) shows that there exists $c > 0$, which depends on Ω, such that

$$\|v\|_{L^2(\Gamma)} \leq c\,\|v\|_{H^1(\Omega)} \quad \forall\, v \in H^1(\Omega). \tag{2.11}$$

The following result can be used to generate various equivalent norms on $H^1(\Omega)$.

Theorem 2.16. *Let Ω be an open, bounded, and connected set in $\mathbb{R}^d$ with a Lipschitz boundary Γ. Assume that $p_1, p_2, \dots, p_N$ are continuous seminorms on $H^1(\Omega)$ with the property that the functional p defined by*

$$p(v) = \|\nabla v\|_{L^2(\Omega)^d} + \sum_{j=1}^{N} p_j(v) \quad \forall v \in H^1(\Omega) \tag{2.12}$$

is a norm on $H^1(\Omega)$. Then p and $\|\cdot\|_{H^1(\Omega)}$ are equivalent norms on $H^1(\Omega)$.

Proof. Since $p_1, p_2, \dots, p_N$ are continuous seminorms on $H^1(\Omega)$ there exist positive constants $M_1, M_2 \dots, M_N$ such that

$$p_j(v) \le M_j \|v\|_{H^1(\Omega)} \quad \forall v \in H^1(\Omega),\ j = 1, \dots N$$

and therefore (2.12) implies that

$$p(v) \le c_1 \|v\|_{H^1(\Omega)} \quad \forall v \in H^1(\Omega), \tag{2.13}$$

where $c_1 = \max\{1, M_1, \dots, M_N\} > 0$.

We shall prove that there exists $c_2 > 0$ such that

$$p(v) \ge c_2 \|v\|_{H^1(\Omega)} \quad \forall v \in H^1(\Omega). \tag{2.14}$$

To this end, we argue by contradiction. Suppose that this inequality is false; then, for all $n \in \mathbb{N}$ there exists $v_n \in H^1(\Omega)$ such that

$$\|v_n\|_{H^1(\Omega)} = 1, \tag{2.15}$$

$$p(v_n) \le \frac{1}{n}. \tag{2.16}$$

We combine (2.12) and (2.16) to see that

$$\|\nabla v_n\|_{L^2(\Omega)^d} + \sum_{j=1}^{N} p_j(v_n) \le \frac{1}{n} \quad \forall n \in \mathbb{N},$$

which shows that $\nabla v_n \to \mathbf{0}$ in $L^2(\Omega)^d$ and $p_j(v_n) \to 0$ for all $j = 1, \dots, N$, as $n \to \infty$. Equality (2.15) shows that $\{v_n\}$ is a bounded sequence in $H^1(\Omega)$ and, using the compact embedding (2.7), it follows that there exists a function $v \in L^2(\Omega)$ and a subsequence of the sequence $\{v_n\}$, still denoted by $\{v_n\}$, such that $v_n \to v$ in $L^2(\Omega)$ as $n \to \infty$. Since

$$\int_\Omega v_n(\boldsymbol{x})\, \partial_i \phi(\boldsymbol{x})\, dx = -\int \partial_i v_n(\boldsymbol{x})\, \phi(\boldsymbol{x})\, dx \quad \forall \phi \in C_0^\infty(\Omega),$$

$v_n \to v$ in $L^2(\Omega)$ and $\partial_i v_n \to 0$ in $L^2(\Omega)$ $\forall\, i = 1, \ldots, d$, passing to the limit in the previous equality we obtain that

$$\int_\Omega v(\boldsymbol{x})\, \partial_i \phi(\boldsymbol{x})\, dx = 0 \quad \forall\, \phi \in C_0^\infty(\Omega) \quad \forall\, i = 1, \ldots, d.$$

This last equality shows that v has weak derivatives of first order and these derivatives vanish a.e. in Ω. We conclude that $v \in H^1(\Omega)$ and $\nabla v = \mathbf{0}$. It follows from (2.10) that

$$\|v_n - v\|_{H^1(\Omega)} = \left(\|v_n - v\|^2_{L^2(\Omega)} + \|\nabla v_n\|^2_{L^2(\Omega)^d}\right)^{\frac{1}{2}}$$

and since $v_n \to v \in L^2(\Omega)$, $\nabla v_n \to \mathbf{0}$ in $L^2(\Omega)^d$, we find that

$$v_n \to v \qquad \text{in} \quad H^1(\Omega) \quad \text{as} \quad n \to \infty. \tag{2.17}$$

Next, since the seminorms $p_1, \ldots, p_N$ are continuous, (2.12) and (2.17) imply that

$$p(v_n) \to p(v) \qquad \text{as } n \to \infty$$

and, therefore, by (2.16) we find that $p(v) = 0$. Using now the fact that p is a norm on the space $H^1(\Omega)$, it follows that $v = 0$, which contradicts the equality $\|v\|_{H^1(\Omega)} = 1$ obtained by combining (2.15) and (2.17). Therefore, inequality (2.14) holds.

It follows that the norm p satisfies (2.13) and (2.14), which concludes the proof. □

Many useful inequalities can be derived as consequence of Theorem 2.16, see [57] and [126] for details. Here, we present only one consequence of the theorem, which will be used in Section 8.2 of this manuscript.

Corollary 2.17. *Let Ω be an open, bounded, and connected set in $\mathbb{R}^d$ with a Lipschitz boundary Γ and let Γ_1 be a measurable part of Γ such that meas $\Gamma_1 > 0$. Then the functional p on $H^1(\Omega)$ defined by*

$$p(v) = \|\nabla v\|_{L^2(\Omega)^d} + \int_{\Gamma_1} |v|\, da \quad \forall\, v \in H^1(\Omega) \tag{2.18}$$

is a norm on $H^1(\Omega)$, equivalent to the norm $\|\cdot\|_{H^1(\Omega)}$.

Proof. We apply Theorem 2.16 to the special case $N = 1$ and

$$p_1(v) = \int_{\Gamma_1} |v|\, da.$$

Clearly, p_1 is a seminorm on the space $H^1(\Omega)$; moreover, by using the Cauchy-Schwarz inequality and (2.11), we have

$$\begin{aligned} p_1(v) = \int_{\Gamma_1} |v|\, da &\le \left(\int_{\Gamma_1} da\right)^{\frac{1}{2}} \left(\int_{\Gamma_1} v^2\, da\right)^{\frac{1}{2}} \\ &\le (meas\, \Gamma_1)^{\frac{1}{2}} \|v\|_{L^2(\Gamma_1)} \le c\,(meas\, \Gamma_1)^{\frac{1}{2}} \|v\|_{H^1(\Omega)} \quad \forall\, v \in H^1(\Omega), \end{aligned}$$

which shows that p_1 is continuous. Assume now that $p(v) = 0$; then $\nabla v = \mathbf{0}$ and $\int_{\Gamma_1} v\, da = 0$; the first equality shows that v is a constant in Ω, $v(\boldsymbol{x}) = k$ a.e. $\boldsymbol{x} \in \Omega$, and the second one shows that

$$\int_{\Gamma_1} k\, da = k\,(meas\, \Gamma_1) = 0,$$

which implies that $k = 0$. Thus, $v = 0$ and therefore p is a norm on $H^1(\Omega)$. Corollary 2.17 is now a consequence of Theorem 2.16. □

2.4 Spaces of Vector-valued Functions

We shall need the spaces of vector-valued functions in studying time-dependent variational problems. In the following, if it is not specified otherwise, $(X, \|\cdot\|_X)$ will denote a real Banach space and $[0,T]$ will denote the time interval of interest, for $T > 0$.

$C^m([0,T];X)$ spaces. We define $C([0,T];X)$ to be the space of functions $v : [0,T] \to X$ that are continuous on the closed interval $[0,T]$. With the norm

$$\|v\|_{C([0,T];X)} = \max_{t\in[0,T]} \|v(t)\|_X,$$

the space $C([0,T];X)$ is a Banach space.

Definition 2.18. A function $v : [0,T] \to X$ is said to be *(strongly) differentiable* at $t_0 \in [0,T]$ if there exists an element in X, denoted as $v'(t_0)$ and called the *(strong) derivative* of v at t_0, such that

$$\lim_{h\to 0} \left\| \frac{1}{h}(v(t_0 + h) - v(t_0)) - v'(t_0) \right\|_X = 0,$$

where the limit is taken with respect to h with $t_0 + h \in [0,T]$. The derivative at $t_0 = 0$ is defined as a right-sided limit and that at $t_0 = T$ as a left-sided limit. The function v is said to be *differentiable on* $[0,T]$ if it is differentiable at every $t_0 \in [0,T]$. It is *differentiable a.e.* if it is differentiable a.e. on $(0,T)$. In this case, the function v' is called the (strong) derivative of v. Higher derivatives $v^{(j)}$, $j \geq 2$, are defined recursively by $v^{(j)} = (v^{(j-1)})'$. Usually we use the notation $\dot{v} = v'$ and we understand $v^{(0)}$ to be v.

For an integer $m \geq 0$, we define the space

$$C^m([0,T];X) = \{\, v \in C([0,T];X) : v^{(j)} \in C([0,T];X),\ j = 1,\dots,m \,\}.$$

This is a Banach space with the norm

$$\|v\|_{C^m([0,T];X)} = \sum_{j=0}^{m} \max_{t\in[0,T]} \|v^{(j)}(t)\|_X.$$

In particular, $C^1([0,T];X)$ denotes the space of continuously differentiable functions on $[0,T]$ with values in X. This is a Banach space with the norm

$$\|v\|_{C^1([0,T];X)} = \max_{t\in[0,T]} \|v(t)\|_X + \max_{t\in[0,T]} \|\dot{v}(t)\|_X.$$

We also set

$$\begin{aligned} C^\infty([0,T];X) &= \bigcap_{m=0}^{\infty} C^m([0,T];X) \\ &\equiv \{\, v \in C([0,T];X) : v \in C^m([0,T];X) \quad \forall m \in \mathbb{Z}_+ \}, \end{aligned}$$

the space of infinitely differentiable functions defined on $[0,T]$ with values in X.

The spaces $L^p(0,T;X)$. For $p \in [1,\infty)$, we define $L^p(0,T;X)$ to be the space of all measurable functions $v : [0,T] \to X$ such that $\int_0^T \|v(t)\|_X^p dt < \infty$. With the norm

$$\|v\|_{L^p(0,T;X)} = \left(\int_0^T \|v(t)\|_X^p dt \right)^{1/p},$$

the space $L^p(0,T;X)$ becomes a Banach space. We define $L^\infty(0,T;X)$ to be the space of all measurable functions $v : [0,T] \to X$ such that $t \mapsto \|v(t)\|_X$ is essentially bounded on $[0,T]$. The space $L^\infty(0,T;X)$ is a Banach space with the norm

$$\|v\|_{L^\infty(0,T;X)} = \operatorname*{ess\,sup}_{t\in[0,T]} \|v(t)\|_X.$$

When $(X,(\cdot,\cdot)_X)$ is a Hilbert space, $L^2(0,T;X)$ is also a Hilbert space with the inner product given by

$$(u,v)_{L^2(0,T;X)} = \int_0^T (u(t),v(t))_X \, dt.$$

Also, below we shall use the notation $L^p(0,T)$ for the space $L^p(0,T;\mathbb{R})$.

The following theorem summarizes some basic properties of the spaces $L^p(0,T;X)$.

Theorem 2.19. *Let $p \in [1,\infty]$. Then:*

(a) *$C([0,T];X)$ is dense in $L^p(0,T;X)$ and the embedding is continuous.*
(b) *If $X \hookrightarrow Y$, then $L^p(0,T;X) \hookrightarrow L^q(0,T;Y)$ for $1 \leq q \leq p \leq \infty$.*

The following result of Lebesgue is useful when we localize a global relation.

Theorem 2.20. *Assume that $v \in L^1(0,T;X)$. Then*

$$\lim_{h\to 0} \frac{1}{h}\int_{t_0}^{t_0+h} \|v(t)-v(t_0)\|_X \, dt = 0 \quad \text{for almost all } t_0 \in (0,T).$$

We see that Theorem 2.20 implies that if $v \in L^1(0,T;X)$, then

$$\lim_{h\to 0} \frac{1}{h}\int_{t_0}^{t_0+h} v(t)\, dt = v(t_0) \quad \text{for almost all } t_0 \in (0,T).$$

Here the limit is understood in the sense of the norm of X, that is,

$$\lim_{h\to 0} \left\| \frac{1}{h}\int_{t_0}^{t_0+h} v(t)\, dt - v(t_0)\right\|_X = 0 \quad \text{for almost all } t_0 \in (0,T).$$

The next result describes the dual of the space $L^p(0,T;X)$, in the Hilbertian case.

Theorem 2.21. *Let $p \in [1,\infty)$ and let $q \in (1,\infty]$ be its conjugate. Assume X is a Hilbert space. Then the dual of the space $L^p(0,T;X)$ is $L^q(0,T;X)$.*

The duality pairing between the space $L^p(0,T;X)$ and its dual is given by

$$\langle u', u\rangle = \int_0^T (u'(t), u(t))_X dt.$$

The spaces $W^{k,p}(0,T;X)$. To define the Sobolev spaces of vector-valued functions, we need to define in an appropriate way derivatives with respect to the time variable for functions defined on $[0,T]$ with values in X. The approach is similar to that taken in the case of weak derivatives of real-valued functions, see Definition 2.5; that is, we take as a starting point the elementary integration by parts formula

$$\int_0^T v(t)\phi^{(m)}(t)\, dt = (-1)^m \int_0^T v^{(m)}(t)\phi(t)\, dt$$

which holds for all functions $v \in C^m([0,T];X)$ and $\phi \in C_0^\infty(0,T)$. Here m is a positive integer and $(\cdot)^{(m)} = d^m(\cdot)/dt^m$.

Definition 2.22. A function $v : [0, T] \to X$ is said to be *locally integrable*, $v \in L^1_{\text{loc}}(0, T; X)$, if for every closed interval $B \subset [0, T]$, we have

$$\int_B \|v(t)\|_X dt < \infty.$$

Definition 2.23. Let $v,\, w \in L^1_{\text{loc}}(0, T; X)$. Then w is called a mth *weak derivative* of v if

$$\int_0^T v(t)\, \phi^{(m)}(t)\, dt = (-1)^m \int_0^T w(t)\, \phi(t)\, dt \quad \forall\, \phi \in C_0^\infty(0, T). \tag{2.19}$$

It can be proved that when a weak derivative exists, it is unique. For this reason, if v and w satisfy (2.19) for some positive integer m we write simply $w = v^{(m)}$. The first two weak derivatives are also denoted by $\dot{v}$, $\ddot{v}$ and we note that $v^{(0)} = v$.

For $k \in \mathbb{Z}_+$ and $1 \le p \le \infty$, we introduce the space

$$W^{k,p}(0, T; X) = \{\, v \in L^p(0, T; X) : \|v^{(m)}\|_{L^p(0,T;X)} < \infty \quad \forall\, m \le k\,\}.$$

When $p < \infty$, we define the norm in the space $W^{k,p}(0, T; X)$ by

$$\|v\|_{W^{k,p}(0,T;X)} = \left(\int_0^T \sum_{0 \le m \le k} \|v^{(m)}(t)\|_X^p dt \right)^{1/p}.$$

When $p = \infty$, the norm is defined as

$$\|v\|_{W^{k,\infty}(0,T;X)} = \max_{0 \le m \le k} \operatorname*{ess\,sup}_{t \in [0,T]} \|v^{(m)}(t)\|_X.$$

If X is a Hilbert space and $p = 2$, then the space

$$H^k(0, T; X) \equiv W^{k,2}(0, T; X)$$

is a Hilbert space with the inner product

$$(u, v)_{H^k(0,T;X)} = \sum_{0 \le m \le k} \int_0^T (u^{(m)}(t), v^{(m)}(t))_X\, dt.$$

We have the following result.

Theorem 2.24. *Assume that X is a Banach space and $p \in [1, \infty]$. Then $W^{1,p}(0, T; X) \hookrightarrow C([0, T]; X)$.*

The previous theorem shows that every element $v \in W^{1,p}(0, T; X)$ can be identified with an element, still denoted by v, in the space $C([0, T]; X)$, possibly after a modification on a subset of $[0, T]$ with zero measure. Moreover, there is a positive constant c such that

$$\|v\|_{C([0,T];X)} \leq c\,\|v\|_{W^{1,p}(0,T;X)} \quad \forall\, v \in W^{1,p}(0,T;X).$$

The next result can help us to understand the nature of the functions in the spaces $W^{k,p}(0,T;X)$. For this, we recall the definition of absolute continuity.

Definition 2.25. A function $v : [0,T] \to X$ is *absolutely continuous* if for any $\varepsilon > 0$, there exists a $\delta = \delta(\varepsilon) > 0$ such that $\sum_n \|v(b_n) - v(a_n)\|_X < \varepsilon$ whenever $\{(a_n, b_n)\}_n \subset [0,T]$ is a countable family of pairwisely disjoint intervals with total length $\sum_n (b_n - a_n) < \delta$.

We have the following characterization for the space $W^{k,p}(0,T;X)$.

Theorem 2.26. *Let $k > 0$ be an integer, $p \in [1,\infty]$, and $v \in L^p(0,T;X)$. Then $v \in W^{k,p}(0,T;X)$ if and only if $v(t) = w(t)$ a.e. on $(0,T)$ for some $w \in C^{k-1}([0,T];X)$ such that w, w', ..., $w^{(k-1)}$ are absolutely continuous on $[0,T]$, the k-th strong derivative $w^{(k)}$ exists a.e., and $w^{(k)} \in L^p(0,T;X)$.*

Because of Theorem 2.26, we usually simply say $v \in W^{k,p}(0,T;X)$ iff v, v', ..., $v^{(k-1)}$ are absolutely continuous on $[0,T]$ and $v^{(k)} \in L^p(0,T;X)$.

It is well known that every real-valued absolutely continuous function $v : [0,T] \to \mathbb{R}$ is differentiable a.e. on $(0,T)$ and it can be expressed as the indefinite integral of its derivative. Some simple examples (see for instance [11, p. 15]) show that this fails when v is absolutely continuous from $[0,T]$ to a general Banach space X. However, the following theorem, obtained in [92], holds.

Theorem 2.27. *Let X be a reflexive Banach space and let $v : [0,T] \to X$ be an absolutely continuous function. Then v is a.e. differentiable on $(0,T)$, its derivative $\dot{v}$ belongs to $L^1(0,T;X)$, and*

$$v(t) = v(0) + \int_0^t \dot{v}(s)\,ds \quad \forall\, t \in [0,T]. \tag{2.20}$$

Using Theorems 2.26 and 2.27, we can show the following results, which will be used on several occasions in the rest of the manuscript.

Proposition 2.28. *Let X be a reflexive Banach space and let $v : [0,T] \to X$. Then:*

(a) *v belongs to $W^{1,1}(0,T;X)$ iff v is an absolutely continuous function on $[0,T]$ to X;*
(b) *v belongs to $W^{1,\infty}(0,T;X)$ iff v is a Lipschitz continuous function on $[0,T]$ to X.*

Proposition 2.29. *Let X be a reflexive Banach space and $u \in W^{1,p}(0,T;X)$ for some $p \in [1,\infty]$. Then*

$$\|u(t) - u(s)\|_X \leq \int_s^t \|\dot{u}(\tau)\|_X d\tau, \quad 0 \leq s \leq t \leq T.$$

We also have

$$\|u(t) - u(s)\|_X^p \leq (t-s)^{p-1} \int_s^t \|\dot{u}(\tau)\|_X^p d\tau, \quad 0 \leq s \leq t \leq T,$$

if $p < \infty$, and

$$\|u(t) - u(s)\|_X \leq (t-s)\|\dot{u}\|_{L^\infty(0,T;X)}, \quad 0 \leq s \leq t \leq T,$$

if $p = \infty$.

Proposition 2.30. *Let X be a Hilbert space and let $a : X \times X \to \mathbb{R}$ be a bilinear symmetric continuous and X-elliptic form. Then, for all $u \in W^{1,2}(0,T;X)$, the real-valued function $t \mapsto \frac{1}{2}\, a(u(t),u(t))$ is absolutely continuous on $[0,T]$,*

$$\frac{d}{dt}\left(\frac{1}{2}\, a(u(t),u(t))\right) = a(u(t),\dot{u}(t)) \quad \text{a.e. on } (0,T),$$

and the following equality holds:

$$\frac{1}{2}\, a(u(t),u(t)) = \frac{1}{2}\, a(u(0),u(0)) + \int_0^t a(u(s),\dot{u}(s))\, ds,$$

for all $t \in [0,T]$.

We end this section with some integrability results that will be useful in the study of evolutionary variational inequalities. The following results represent direct consequences of more general results that can be found in [16, p. 160].

Theorem 2.31. *Let X be a Banach space and let $j : X \to (-\infty,\infty]$ be a proper convex l.s.c. function. Then:*

(1) For all $v \in L^1(0,T;X)$ the function $t \mapsto j(v(t))$ is measurable on $[0,T]$. Moreover, it is integrable if and only if there exists $g \in L^1(0,T)$ such that $j(v(t)) \leq g(t)$ a.e. $t \in (0,T)$.

(2) The function $\phi : L^1(0,T;X) \to (-\infty,\infty]$ defined by

$$\phi(v) = \begin{cases} \displaystyle\int_0^T j(v(t))\, dt & \text{if} \quad j(v) \in L^1(0,T;X), \\ \infty & \text{otherwise,} \end{cases}$$

is proper, convex, and lower semicontinuous.

Direct consequences of Theorem 2.31 are provided by the following results.

Corollary 2.32. *Let X be a Banach space and let $j : X \to \mathbb{R}$ be a convex l.s.c. function that satisfies*

$$j(v) \leq c\left(\|v\|_X^2 + 1\right) \quad \forall\, v \in X, \tag{2.21}$$

for some $c > 0$. Then for all $v \in L^2(0,T;X)$, the function $t \mapsto j(v(t))$ is integrable on $[0,T]$. Moreover,

$$v \mapsto \phi(v) = \int_0^T j(v(t))\, dt \tag{2.22}$$

is a convex l.s.c. function on the space $L^2(0,T;X)$.

Proof. Let $v \in L^2(0,T;X)$ and take $g(t) = c\left(\|v(t)\|_X^2 + 1\right)$ a.e. $t \in (0,T)$. It follows that $g \in L^1(0,T)$. We use now (2.21) and Theorem 2.31(1) to conclude the first part of the corollary. The second part is a consequence of Theorem 2.31(2), combined with the inclusion $L^2(0,T;X) \hookrightarrow L^1(0,T;X)$ provided by Theorem 2.19. □

Corollary 2.33. *Let $(X, (\cdot,\cdot)_X)$ be a Hilbert space and let $j : X \to \mathbb{R}$ be a convex Gâteaux differentiable function that satisfies*

$$\|\nabla j(v)\|_X \leq c\left(\|v\|_X + 1\right) \quad \forall\, v \in X, \tag{2.23}$$

for some $c > 0$. Then for all $v \in L^2(0,T;X)$ the function $t \mapsto j(v(t))$ is integrable on $[0,T]$. Moreover, the function $\phi : L^2(0,T;X) \to \mathbb{R}$ defined by (2.22) is convex and l.s.c.

Proof. We use Corollary 1.37 to see that j is lower semicontinuous. Moreover, it follows from Proposition 1.36 that

$$j(w) - j(v) \geq (\nabla j(v), w - v)_X \quad \forall\, v,\, w \in X.$$

We choose $w = 0_X$ in this inequality and use the Cauchy-Schwarz inequality to find that

$$j(v) \leq \|\nabla j(v)\|_X \|v\|_X + j(0_X) \quad \forall\, v \in X. \tag{2.24}$$

We combine inequalities (2.24) and (2.23) to see that j satisfies an inequality of the form (2.21), then we use Corollary 2.32 to conclude the proof. □

Bibliographical Notes

The material presented in Chapter 1 is standard and can be found in many books on functional analysis. For more information in the field, we refer the reader to the books [7], [18], [150]–[156]. A complete treatment of the general theory of convex functions as well as proofs of the results on convex analysis presented in Section 1.4 can be found in the works [13, 38, 44, 66, 151], for instance.

For a comprehensive treatment of basic aspects of spaces introduced in Sections 2.1 and 2.2, we refer the reader to [1, 18, 95, 108, 121]. The proof of Theorem 2.16 was written following [126, p. 201]. A more general result that can be used to generate various equivalent norms on the Sobolev space $W^{k,p}(\Omega)$ can be found in [57, p. 117]. More details on the spaces of vector-valued functions presented in Section 2.4 can be found in [11, 13, 17, 22].

Part II
Variational Inequalities

Chapter 3
Elliptic Variational Inequalities

In this chapter, we present some theorems on the solvability of elliptic variational and quasivariational inequalities. We start with a basic existence and uniqueness result for elliptic variational inequalities, then we provide convergence results. Next, we extend part of these results to the study of elliptic quasivariational and time-dependent variational and quasivariational inequalities, respectively. The results presented in this chapter will be applied in the study of static antiplane frictional contact problems with elastic materials. They also are crucial tools in deriving existence results for evolutionary variational inequalities. Everywhere in this chapter, X denotes a real Hilbert space with inner product $(\cdot,\cdot)_X$ and norm $\|\cdot\|_X$.

3.1 A Basic Existence and Uniqueness Result

Let $a : X \times X \to \mathbb{R}$ and let $j : X \to (-\infty, \infty]$. Given $f \in X$, we consider the problem of finding an element $u \in X$ such that

$$a(u, v - u) + j(v) - j(u) \geq (f, v - u)_X \quad \forall\, v \in X. \tag{3.1}$$

An inequality of the form (3.1) is called an *elliptic variational inequality of the second kind.* We are interested in sufficient conditions for the existence and uniqueness of a solution to inequality (3.1). To this end, we make the following assumptions:

$$\left.\begin{array}{l} a : X \times X \to \mathbb{R} \text{ is a bilinear symmetric form and} \\ \text{(a) there exists } M > 0 \text{ such that} \\ \qquad |a(u, v)| \leq M\, \|u\|_X\, \|v\|_X \quad \forall\, u, v \in X. \\ \text{(b) there exists } m > 0 \text{ such that} \\ \qquad a(v, v) \geq m\, \|v\|_X^2 \quad \forall\, v \in X. \end{array}\right\} \tag{3.2}$$

$$j : X \to (-\infty, \infty] \text{ is a proper convex l.s.c. function.} \tag{3.3}$$

Note that according to Definition 1.15, conditions (3.2)(a) and (3.2)(b) show that the bilinear form a is continuous and X-elliptic, respectively.

We have the following basic existence and uniqueness result.

Theorem 3.1. *Let X be a Hilbert space and assume that* (3.2) *and* (3.3) *hold. Then, for each $f \in X$, the elliptic variational inequality* (3.1) *has a unique solution. Moreover, the solution depends Lipschitz continuously on f.*

The proof of Theorem 3.1 will be carried out in three steps, and it is based on arguments for the minimization of convex functions. Below, we assume that (3.2) and (3.3) hold and, for all $f \in X$, we define the function $J_f : X \to (-\infty, \infty]$ by the formula

$$J_f(v) = \frac{1}{2}\, a(v,\, v) + j(v) - (f, v)_X. \tag{3.4}$$

The first step is given by the following equivalence result.

Lemma 3.2. *Let $f \in X$. An element $u \in X$ is solution of the variational inequality* (3.1) *if and only if u is a minimizer of J_f on X, that is,*

$$J_f(v) \geq J_f(u) \quad \forall\, v \in X. \tag{3.5}$$

Proof. Assume that u is a solution of (3.1). Then, since j is a proper function, it follows that $j(u) < \infty$ and therefore the operations below are well-defined. Using the properties of the form a, we have

$$J_f(v) - J_f(u) = a(u, v - u) + j(v) - j(u) - (f, v - u)_X + \frac{1}{2}\, a(u - v, u - v),$$

for every $v \in X$. Therefore, by using (3.1), (3.2)(b) we obtain

$$J_f(v) - J_f(u) \geq 0 \quad \forall\, v \in X,$$

which shows that u is a solution of the minimization problem (3.5).

Conversely, assume that (3.5) holds and let $v \in X$, $t \in (0, 1)$. We have

$$J_f(u + t(v - u)) \geq J_f(u)$$

and, using the definition (3.4), the convexity of j and the properties of a, we deduce that

$$t\, a(u, v - u) + \frac{t^2}{2}\, a(u - v, u - v) + t(j(v) - j(u)) \geq t\, (f, v - u)_X.$$

We divide both sides of the previous inequality by t and pass to the limit as $t \to 0^+$ to conclude that u satisfies the variational inequality (3.1). □

In the second step, we have the following existence and uniqueness result.

Lemma 3.3. *Let $f \in X$. Then there exists a unique element $u \in X$ such that (3.5) hold.*

Proof. Let $l_f = \inf_{v \in X} J_f(v)$. Since $j : X \to (-\infty, \infty]$ is a proper function, we deduce that $J_f : X \to (-\infty, \infty]$ is a proper function and, therefore,

$$l_f < \infty. \tag{3.6}$$

Consider $\{u_n\} \subset X$ a sequence of elements of X such that

$$J_f(u_n) \to l_f \quad \text{as } n \to \infty. \tag{3.7}$$

It follows from Proposition 1.34 that there exists $\alpha \in X$ and $\beta \in \mathbb{R}$ such that $j(u_n) \geq (\alpha, u_n)_X + \beta$ for all $n \in \mathbb{N}$. Using this inequality and (3.2)(b), we obtain

$$J_f(u_n) \geq \frac{m}{2} \|u_n\|_X^2 - \|\alpha\|_X \|u_n\|_X - \|f\|_X \|u_n\|_X + \beta \quad \forall\, n \in \mathbb{N}. \tag{3.8}$$

This shows that $\{u_n\}$ is a bounded sequence in X. Indeed, in the opposite case there exists a subsequence $\{u_{n_k}\} \subset \{u_n\}$ such that $\|u_{n_k}\|_X \to \infty$ as $k \to \infty$, and (3.8) implies that

$$J_f(u_{n_k}) \to \infty \quad \text{as } k \to \infty. \tag{3.9}$$

We combine now (3.7) and (3.9) to see that $l_f = \infty$, which contradicts (3.6).

It follows now from Theorem 1.21 that there exists an element $u \in X$ and a subsequence $\{u_{n_p}\} \subset \{u_n\}$ such that

$$u_{n_p} \rightharpoonup u \text{ in } X \quad \text{as } p \to \infty. \tag{3.10}$$

Also, it follows from Proposition 1.33 that J_f is a convex lower semicontinuous function on X and, therefore, (3.10) implies that

$$\liminf_{p \to \infty} J_f(u_{n_p}) \geq J_f(u). \tag{3.11}$$

Combining (3.7) and (3.11), we deduce that $l_f \geq J_f(u)$. The reverse inequality follows from the definition of l_f. Therefore, we conclude that $J_f(u) = l_f$, which shows that the element $u \in X$ satisfies (3.5).

The uniqueness of the solution of the minimization problem (3.5) follows from assumptions (3.2) and (3.3). Indeed, let $u_1 \neq u_2$ be two elements of X that satisfy (3.5); a simple computation shows that

$$\begin{aligned}
&\frac{1}{2} J_f(u_1) + \frac{1}{2} J_f(u_2) - J_f\Big(\frac{u_1 + u_2}{2}\Big) \\
&\quad = \frac{1}{8}\, a(u_1 - u_2, u_1 - u_2) + \frac{1}{2}\, j(u_1) + \frac{1}{2}\, j(u_2) - j\Big(\frac{u_1 + u_2}{2}\Big)
\end{aligned}$$

and, using (3.2)(b) and the convexity of the function j, since $u_1 \neq u_2$ we find that

$$\frac{1}{2}\, J_f(u_1) + \frac{1}{2}\, J_f(u_2) > J_f\Big(\frac{u_1+u_2}{2}\Big).$$

On the other hand $J_f(u_1) = J_f(u_2) = \inf_{v\in X} J_f(v)$ and, therefore, the previous inequality yields

$$\inf_{v\in X} J_f(v) > J_f\Big(\frac{u_1+u_2}{2}\Big)$$

which is a contradiction. We conclude that $u_1 = u_2$, and this completes the proof of the lemma. □

We now conclude the proof of Theorem 3.1.

Proof. The existence and uniqueness part in Theorem 3.1 follows as a direct consequence of Lemmas 3.2 and 3.3. Assume now that u_1 and u_2 are the solutions of the inequality (3.1) for $f = f_1$ and $f = f_2$, respectively. Then, $j(u_1) < \infty$, $j(u_2) < \infty$ and, moreover,

$$a(u_1, v-u_1) + j(v) - j(u_1) \geq (f_1, v-u_1)_X \quad \forall\, v \in X,$$
$$a(u_2, v-u_2) + j(v) - j(u_2) \geq (f_2, v-u_2)_X \quad \forall\, v \in X.$$

We choose now $v = u_2$ in the first inequality, $v = u_1$ in the second one, and add the resulting inequalities to obtain

$$a(u_1-u_2, u_1-u_2) \leq (f_1-f_2, u_1-u_2)_X.$$

Using again (3.2)(b) and the Cauchy-Schwarz inequality, we find

$$\|u_1-u_2\|_X \leq \frac{1}{m}\, \|f_1-f_2\|_X \tag{3.12}$$

which shows that the solution of (3.1) depends Lipschitz continuously on f, with Lipschitz constant $L \leq \frac{1}{m}$. □

We can apply Theorem 3.1 to the case when j is the indicator function of a nonempty, convex, and closed subset $K \subset X$, (1.6). It follows that ψ_K is a proper, convex, and l.s.c. function and therefore, by taking $j = \psi_K$ in Theorem 3.1, we obtain the following result.

Corollary 3.4. *Let X be a Hilbert space, let $K \subset X$ be a nonempty, convex, and closed subset, and assume* (3.2). *Then for each $f \in X$, there exists a unique element $u \in K$ such that*

$$a(u, v-u) \geq (f, v-u)_X \quad \forall\, v \in K. \tag{3.13}$$

Moreover, the solution depends Lipschitz continuously on f.

A problem of the form (3.13) is called *elliptic variational inequality of the first kind.* From the historical point of view, elliptic variational inequalities of the first kind were studied first and constituted a point of departure for introduction and study of elliptic variational inequalities of the second kind. A classical exemple is provided by the obstacle problem, see for instance [57, p. 138]. A second example can be obtained by choosing $a(u, v) = (u, v)_X$ for all $u, v \in X$; then, by using (1.3), it is easy to see that in this case the solution of (3.13) is the projection of the element f on K, i.e., $u = P_K f$.

Elliptic variational inequality of the first kind arises in the study of unilateral contact between a linearly elastic body and a rigid foundation, see for instance [42, 86]. In this book, we do not present results concerning such contact models, since they are not relevant to the study of antiplane frictional contact problems, which are the aim of this monograph. And, for this reason, we deal only with elliptic variational inequalities of the second kind, which we call, for simplicity, *elliptic variational inequalities.*

Next, by using $j \equiv 0$ in Theorem 3.1 or $K = X$ in Corollary 3.4, we obtain the following version of the well-known Lax-Milgram theorem.

Corollary 3.5. *Let X be a Hilbert space and assume that* (3.2) *holds. Then, for each $f \in X$, there exists a unique element $u \in X$ such that*

$$a(u, v) = (f, v)_X \quad \forall\, v \in X. \tag{3.14}$$

Moreover, the solution depends linearly and continuously on f.

A problem of the form (3.14) is called *variational equation.* Such kind of problems arise in the study of linear elliptic boundary value problems, see for instance [7, 18].

We end this section with the remark that an element $u \in K$ is a solution of problem (3.13) if and only if

$$J_f(v) \geq J_f(u) \quad \forall\, v \in K$$

where $J_f : X \to \mathbb{R}$ is the functional defined by

$$J_f(v) = \frac{1}{2}\, a(v, v) - (f, v)_X. \tag{3.15}$$

Also, an element $u \in X$ is a solution of problem (3.14) if and only if

$$J_f(v) \geq J_f(u) \quad \forall\, v \in X,$$

where, again, J_f is given by (3.15). The statements above represent a direct consequence of Lemma 3.2, combined with definitions (3.4) and (1.6).

3.2 Convergence Results

In this section, we study the dependence of the solution of the variational inequality (3.1) on the function j. To this end, we assume in what follows that (3.2) and (3.3) hold, $f \in X$ and, for every $\rho > 0$, let j_ρ be a perturbation of j that satisfies (3.3). We consider the problem of finding an element $u_\rho \in X$ such that

$$a(u_\rho, v - u_\rho) + j_\rho(v) - j_\rho(u_\rho) \geq (f, v - u_\rho)_X \quad \forall\, v \in X. \tag{3.16}$$

We deduce from Theorem 3.1 that (3.1) has a unique solution $u \in X$ and, for each $\rho > 0$, (3.16) has a unique solution $u_\rho \in X$. Moreover, since j and j_ρ satisfy condition (3.3), it follows that $j(u) < \infty$ and $j_\rho(u_\rho) < \infty$. In practice, the interest is to consider a function j_ρ that is more regular than j (for instance, which is Gâteaux differentiable) and, for this reason, the variational inequality (3.16) may be considered a *regularization* of the variational inequality (3.1).

Consider now the following assumptions.

$$\left.\begin{array}{l} \text{There exists } F : \mathbb{R}_+ \to \mathbb{R}_+ \ \text{ such that} \\ \text{(a) } |j_\rho(v) - j(v)| \leq F(\rho) \quad \forall\, v \in X, \ \text{for each } \rho > 0. \\ \text{(b) } \lim\limits_{\rho \to 0} F(\rho) = 0. \end{array}\right\} \tag{3.17}$$

$$\left.\begin{array}{l} \text{(a) } j_\rho(v) \geq 0 \quad \forall\, v \in X \ \text{ and } \ j_\rho(0_X) = 0, \ \text{for each } \rho > 0. \\ \text{(b) } j_\rho(v) \to j(v) \ \text{as } \rho \to 0, \ \forall\, v \in X. \\ \text{(c) For each sequence } \{v_\rho\} \subset X \ \text{ such that } v_\rho \rightharpoonup v \in X \\ \qquad \text{as } \rho \to 0 \text{ one has } \liminf\limits_{\rho \to 0} j_\rho(v_\rho) \geq j(v). \end{array}\right\} \tag{3.18}$$

Note that condition (3.18)(b) above is understood in the following sense: for every sequence $\{\rho_n\} \subset \mathbb{R}$ converging to 0 as $n \to \infty$, one has $j_{\rho_n}(v) \to j(v)$ as $n \to \infty$, for all $v \in X$. We shall use such notation to indicate various convergences or inequalities everywhere in the book, see for instance assumption (3.18)(c) above and the convergence (3.19) below.

The behavior of the solution u_ρ as ρ converges to zero is given in the following theorem.

Theorem 3.6. *Under the assumptions* (3.17) *or* (3.18), *the solution* u_ρ *of problem* (3.16) *converges to the solution* u *of problem* (3.1), *i.e.,*

$$u_\rho \to u \quad \text{in } X \text{ as } \rho \to 0. \tag{3.19}$$

Proof. Let $\rho > 0$ and assume that (3.17) holds. We take $v = u_\rho$ in (3.1) and $v = u$ in (3.16) and add the corresponding inequalities to obtain

$$\begin{aligned} a(u_\rho - u, u_\rho - u) &\leq j(u_\rho) - j(u) + j_\rho(u) - j_\rho(u_\rho) \\ &\leq |j_\rho(u_\rho) - j(u_\rho)| + |j_\rho(u) - j(u)|. \end{aligned}$$

We use now the X-ellipticity of the form a, (3.2)(b), and the assumption (3.17)(a) on j_ρ to find that

$$m \, \|u_\rho - u\|_X^2 \leq 2 \, F(\rho). \tag{3.20}$$

The convergence result (3.19) is now a consequence of the inequality (3.20) combined with the assumption (3.17)(b).

Assume now that (3.18) holds and let $\rho > 0$. We choose $v = 0_X$ in (3.16) and use assumption (3.18)(a) to find that

$$a(u_\rho, \, u_\rho) \leq (f, \, u_\rho)_X.$$

It follows now from (3.2)(b) that

$$\|u_\rho\|_X \leq \frac{1}{m} \, \|f\|_X,$$

which shows that $\{u_\rho\}$ is a bounded sequence in X. Therefore, by Theorem 1.21 it follows that there exists a subsequence $\{u_{\rho'}\}$ of $\{u_\rho\}$ and an element $\tilde{u} \in X$ such that

$$u_{\rho'} \rightharpoonup \tilde{u} \text{ in } X \text{ as } \rho' \to 0. \tag{3.21}$$

We use (3.16) to obtain

$$a(u_{\rho'}, v) + j_{\rho'}(v) \geq (f, v - u_{\rho'})_X + a(u_{\rho'}, u_{\rho'}) + j_{\rho'}(u_{\rho'}) \quad \forall \, v \in X,$$

then we pass to the lower limit as $\rho' \to 0$ and use (3.21), Proposition 1.33, and assumption (3.18)(b)(c); as a result we find that

$$a(\tilde{u}, v) + j(v) \geq (f, v - \tilde{u})_X + a(\tilde{u}, \tilde{u}) + j(\tilde{u}) \quad \forall \, v \in X.$$

The previous inequality shows that $\tilde{u}$ satisfies (3.1) and, since by Theorem 3.1 this last inequality has a unique solution, we deduce that

$$\tilde{u} = u. \tag{3.22}$$

We combine now (3.21) and (3.22) to see that every weakly convergent subsequence of the sequence $\{u_\rho\}$ converges to u and therefore, as noted on page 8, the whole sequence $\{u_\rho\}$ converges weakly to u,

$$u_\rho \rightharpoonup u \text{ in } X \text{ as } \rho \to 0. \tag{3.23}$$

We now use the weak convergence (3.23) and inequality (3.16) to prove the strong convergence (3.19). To this end, we consider $\rho > 0$ and take $v = u$ in (3.16) to obtain

$$a(u_\rho, u_\rho - u) \le (f, u_\rho - u)_X + j_\rho(u) - j_\rho(u_\rho),$$

which implies that

$$a(u_\rho - u, u_\rho - u) \le (f, u_\rho - u)_X + j_\rho(u) - j_\rho(u_\rho) + a(u, u - u_\rho).$$

We now use the X-ellipticity of a, (3.2)(b), to find that

$$m \, \|u_\rho - u\|_X^2 \le (f, u_\rho - u)_X + j_\rho(u) - j_\rho(u_\rho) + a(u, u - u_\rho). \tag{3.24}$$

It follows from (3.18)(b) that

$$\begin{aligned} \limsup_{\rho \to 0} \big(j_\rho(u) - j_\rho(u_\rho)\big) &= \limsup_{\rho \to 0} \big(j_\rho(u) - j(u) + j(u) - j_\rho(u_\rho)\big) \\ &= \limsup_{\rho \to 0} \big(j(u) - j_\rho(u_\rho)\big) = j(u) - \liminf_{\rho \to 0} j_\rho(u_\rho) \end{aligned}$$

and, using (3.23) combined with (3.18)(c), we find that

$$\limsup_{\rho \to 0} \big(j_\rho(u) - j_\rho(u_\rho)\big) \le 0. \tag{3.25}$$

We now use (3.24), (3.25), and (3.23) to see that

$$\limsup_{\rho \to 0} \|u_\rho - u\|_X^2 \le 0$$

and, therefore, we conclude that (3.19) holds. □

Theorem 3.6 shows that under conditions (3.17) or (3.18), the solution u of the variational inequality (3.1) can be approached by the solution of the variational inequality (3.16) in the sense of (3.19). In addition to the interest from the asymptotic analysis point of view, this result is important from the numerical point of view. Indeed, in the case when j_ρ is a Gâteaux differentiable function, solving (3.16) is equivalent to solving the nonlinear equation

$$A u_\rho + \nabla j_\rho(u_\rho) = f \tag{3.26}$$

where $A : X \to X$ is the linear continuous operator given by

$$(Au, v)_X = a(u, v) \quad \forall\, u,\, v \in X \tag{3.27}$$

and $\nabla j_\rho : X \to X$ is the gradient operator of j_ρ, i.e.,

$$(\nabla j_\rho(u),\, v)_X = \lim_{h \to 0} \frac{j_\rho(u + hv) - j_\rho(u)}{h} \qquad \forall\, u,\, v \in X. \tag{3.28}$$

We conclude that the solution of the variational inequality (3.1) leads to the solution of the nonlinear equation (3.26) when the parameter ρ is "small" and, to this end, various numerical methods can be used, see, e.g., [52].

3.3 Elliptic Quasivariational Inequalities

In this section, we consider *elliptic quasivariational inequalities*, i.e., we allow the functional j to depend explicitly on the solution. Thus, for a given $f \in X$, we consider the problem of finding an element $u \in X$ such that

$$a(u, v-u) + j(u, v) - j(u, u) \geq (f, v-u)_X \quad \forall\, v \in X. \tag{3.29}$$

The main ideas on the solvability of (3.29) are based on fixed point arguments. To illustrate them, we start with the Banach fixed point argument and, to this end, we consider the following assumption:

$$\left.\begin{array}{l} j : X \times X \to \mathbb{R} \text{ and} \\ \text{(a) for all } \eta \in X,\ j(\eta, \cdot) \text{ is convex and l.s.c. on } X. \\ \text{(b) there exists } \alpha \geq 0 \text{ such that} \\ \qquad j(\eta_1, v_2) - j(\eta_1, v_1) + j(\eta_2, v_1) - j(\eta_2, v_2) \\ \qquad\quad \leq \alpha\, \|\eta_1 - \eta_2\|_X \|v_1 - v_2\|_X \quad \forall\, \eta_1, \eta_2, v_1, v_2 \in X. \end{array}\right\} \tag{3.30}$$

We have the following existence and uniqueness result.

Theorem 3.7. *Let X be a Hilbert space and assume that* (3.2) *and* (3.30) *hold. Assume, moreover, that $m > \alpha$. Then, for each $f \in X$, the quasivariational elliptic inequality* (3.29) *has a unique solution that depends Lipschitz continuously on f.*

The proof of Theorem 3.7 will be carried out in three steps. We assume in what follows that (3.2) and (3.30) hold and let $f \in X$. In the first step, for every $\eta \in X$ we consider the auxiliary problem of finding $u_\eta \in X$, which solves the elliptic variational inequality

$$a(u_\eta, v - u_\eta) + j(\eta, v) - j(\eta, u_\eta) \geq (f, v - u_\eta)_X \quad \forall\, v \in X. \tag{3.31}$$

We use Theorem 3.1 to derive the following existence and uniqueness result.

Lemma 3.8. *For each $\eta \in X$, there exists a unique solution u_η of* (3.31).

We use Lemma 3.8 to define the operator $\Lambda : X \to X$ by

$$\Lambda\eta = u_\eta \quad \forall\, \eta \in X \tag{3.32}$$

and we continue with the following fixed point result.

Lemma 3.9. *If $m > \alpha$, then Λ has a unique fixed point $\eta^* \in X$.*

Proof. Let $\eta_1,, \eta_2 \in X$ and let u_i denote the solution of (3.31) for $\eta = \eta_i$, i.e., $u_i = u_{\eta_i}$, $i = 1, 2$. We have

$$\begin{aligned} a(u_1, v - u_1) + j(\eta_1, v) - j(\eta_1, u_1) &\geq (f, v - u_1)_X \quad \forall\, v \in X, \\ a(u_2, v - u_2) + j(\eta_2, v) - j(\eta_2, u_2) &\geq (f, v - u_2)_X \quad \forall\, v \in X. \end{aligned}$$

We take $v = u_2$ in the first inequality, $v = u_1$ in the second one, and add the resulting inequalities to obtain

$$a(u_1 - u_2, u_1 - u_2) \leq j(\eta_1, u_2) - j(\eta_1, u_1) + j(\eta_2, u_1) - j(\eta_2, u_2). \tag{3.33}$$

We use now (3.2)(b) and (3.30)(b) in (3.33) to see that

$$\|u_1 - u_2\|_X \leq \frac{\alpha}{m}\, \|\eta_1 - \eta_2\|_X. \tag{3.34}$$

Since $m > \alpha$, the inequality (3.34) shows that the operator Λ given by (3.32) is a contraction on the space X and therefore Lemma 3.9 follows from Theorem 1.40. □

We have now all the ingredients to provide the proof of Theorem 3.7.

Proof. Let η^* be the fixed point of the operator Λ obtained in Lemma 3.9. Since $\eta^* = \Lambda\eta^* = u_{\eta^*}$, it follows from (3.31) that u_{η^*} is a solution of the quasivariational inequality (3.29), which concludes the existence part.

Now, let u be a solution of (3.29) and denote $\eta = u$. It follows that u is a solution of the variational inequality (3.31) and, since by Lemma 3.8 this inequality has a unique solution denoted u_η, we have $u = u_\eta$. Moreover, it follows that $\eta = u_\eta$ and, keeping in mind the definition (3.32) of the operator Λ, we deduce that $\eta = \Lambda\eta$. Since by Lemma 3.9 the operator Λ has a unique fixed point, denoted by η^*, we find that $\eta = \eta^*$, which shows that $u = u_{\eta^*}$, and this concludes the proof of the uniqueness.

Assume now that u_1 and u_2 are solutions of the inequality (3.29) for $f = f_1$ and $f = f_2$, respectively. Using arguments similar to those used in the proof of (3.33), we find that

$$\begin{aligned} a(u_1 - u_2, u_1 - u_2) \leq\ & j(u_1, u_2) - j(u_1, u_1) + j(u_2, u_1) - j(u_2, u_2) \\ & + (f_1 - f_2, u_1 - u_2)_X. \end{aligned}$$

We use (3.2)(b) and (3.30)(b) in the previous inequality to see that

$$(m - \alpha)\|u_1 - u_2\|_X \leq \|f_1 - f_2\|_X$$

and, since $m > \alpha$, we find that

$$\|u_1 - u_2\|_X \leq \frac{1}{m - \alpha}\, \|f_1 - f_2\|_X. \tag{3.35}$$

Inequality (3.35) shows that the solution of (3.29) depends Lipschitz continuously on f, which concludes the proof. □

We now investigate the quasivariational inequality (3.29) by using the Schauder-Tychonoff fixed point argument. To this end, we assume in what follows that the functional $j : X \times X \to \mathbb{R}$ satisfies:

$$\text{For all } \eta \in X,\ j(\eta,\,\cdot) \text{ is a seminorm on } X. \tag{3.36}$$

$$\left.\begin{array}{l}\text{For all sequences } \{\eta_n\} \subset X \text{ and } \{u_n\} \subset X \text{ such that}\\ \eta_n \rightharpoonup \eta \in X,\ u_n \rightharpoonup u \in X \text{ and for every } v \in X,\\ \text{the inequality below holds:}\\ \limsup\limits_{n\to\infty} \left[j(\eta_n, v) - j(\eta_n, u_n)\right] \le j(\eta, v) - j(\eta, u).\end{array}\right\} \tag{3.37}$$

$$\left.\begin{array}{l}j(u,v) - j(u,u) + j(v,u) - j(v,v) < m\,\|u-v\|_X^2\\ \forall\, u,\, v \in X,\, u \neq v.\end{array}\right\} \tag{3.38}$$

$$\left.\begin{array}{l}\text{There exists } \alpha \in (0,m) \text{ such that}\\ j(u,v) - j(u,u) + j(v,u) - j(v,v) \le \alpha\,\|u-v\|_X^2\\ \forall\, u,\, v \in X.\end{array}\right\} \tag{3.39}$$

Our second result in this section is the following.

Theorem 3.10. *Let X be a separable Hilbert space, $f \in X$, and assume that* (3.2) *holds. Then:*

(1) Under the assumptions (3.36) *and* (3.37), *there exists at least one solution to the quasivariational inequality* (3.29).

(2) Under the assumptions (3.36), (3.37), *and* (3.38), *there exists a unique solution to the quasivariational inequality* (3.29).

(3) Under the assumptions (3.36), (3.37), *and* (3.39), *there exists a unique solution to the quasivariational inequality* (3.29), *which depends Lipschitz continuously on $f \in X$.*

The proof of Theorem 3.10 will be carried out in three steps. We assume in what follows that (3.2), (3.36), and (3.37) hold and let $f \in X$. As in the proof of Theorem 3.7, we start with the study of the elliptic variational inequality (3.31), defined for all $\eta \in X$. It follows from (3.36) that $j(\eta,\,\cdot) : X \to \mathbb{R}$ is a convex function; moreover, choosing $\eta_n = \eta$ in (3.37) results in

$$\liminf_{n\to\infty}\, j(\eta,\, u_n) \ge j(\eta,\, u)$$

whenever $u_n \rightharpoonup u$ in X, i.e., $j(\eta,\,\cdot)$ is a l.s.c. function on X, for all $\eta \in X$. Then, by using Theorem 3.1 we deduce that Lemma 3.8 holds, and this represents the first step in the proof of Theorem 3.10.

Next, we define the operator $\Lambda : X \to X$ by (3.32) and, in the second step, we have the following result.

Lemma 3.11. *The operator Λ is weakly sequentially continuous, i.e., $\eta_n \rightharpoonup \eta$ in X implies $\Lambda\eta_n \rightharpoonup \Lambda\eta$ in X.*

Proof. Let $\{\eta_n\} \subset X$ such that

$$\eta_n \rightharpoonup \eta \in X \tag{3.40}$$

and denote $\Lambda\eta_n = u_n$ for all $n \in \mathbb{N}$. It follows from (3.31) that

$$a(u_n, v - u_n) + j(\eta_n, v) - j(\eta_n, u_n) \geq (f, v - u_n)_X \quad \forall v \in X. \tag{3.41}$$

Choosing $v = 0_X$ and using (3.36) yields

$$a(u_n, u_n) \leq (f, u_n)_X \quad \forall n \in \mathbb{N}.$$

The previous inequality combined with (3.2) shows that

$$\|u_n\|_X \leq \frac{1}{m}\|f\|_X \quad \forall n \in \mathbb{N}, \tag{3.42}$$

which implies that $\{u_n\}$ is a bounded sequence on X. We deduce by Theorem 1.21 that there exists a subsequence of $\{u_n\}$, again denoted $\{u_n\}$, and an element $w \in X$, such that

$$u_n \rightharpoonup w \text{ in } X \text{ as } n \to \infty. \tag{3.43}$$

On the other hand (3.2) implies that

$$\begin{aligned} a(u_n - w, u_n - w) &= a(u_n, u_n - v) + a(u_n, v - w) - a(w, u_n - w) \\ &\geq 0 \quad \forall v \in X, \ n \in \mathbb{N} \end{aligned}$$

which leads to the inequality

$$a(u_n, u_n - v) \geq a(u_n, w - v) + a(w, u_n - w) \quad \forall v \in X, \ n \in \mathbb{N}.$$

We take the lower limit as $n \to \infty$ in the previous inequality and use (3.43) to obtain

$$\liminf_{n\to\infty} a(u_n, u_n - v) \geq a(w, w - v) \quad \forall v \in X. \tag{3.44}$$

It follows from (3.41) that

$$a(u_n, u_n - v) \leq (f, u_n - v)_X + j(\eta_n, v) - j(\eta_n, u_n) \quad \forall v \in X, \ n \in \mathbb{N}$$

and, using (3.40), (3.43), and (3.37), we obtain

$$\limsup_{n\to\infty} a(u_n, u_n - v) \leq (f, w - v)_X + j(\eta, v) - j(\eta, w) \quad \forall v \in X. \tag{3.45}$$

Combining (3.44) and (3.45), we infer that

$$a(w,\, w - v) \le (f,\, w - v)_X + j(\eta,\, v) - j(\eta,\, w) \quad \forall\, v \in X,$$

and it follows that w solves the elliptic variational inequality (3.31). Since by Lemma 3.8 this last inequality has a unique solution denoted u_η, we deduce that $w = u_\eta$ and, using (3.32), we obtain that

$$\Lambda\eta = w. \tag{3.46}$$

It follows from (3.43) and (3.46) that $\Lambda\eta$ is the unique weak limit of each weakly convergent subsequence of the sequence $\{\Lambda\eta_n\}$ and, therefore, the whole sequence $\{\Lambda\eta_n\}$ converges weakly in X to $\Lambda\eta$. This concludes the proof of the lemma. □

We have now all the ingredients needed to prove the Theorem 3.10.

Proof. (1) Let $f \in X$ and denote $K = \overline{B}(0,\, \frac{1}{m}\, \|f\|_X)$ the closed ball of radius $\frac{1}{m}\|f\|_X$ centered at 0_X, i.e.,

$$K = \left\{ v \in X \,:\, \|v\|_X \le \frac{1}{m}\, \|f\|_X \right\}.$$

Clearly, K is a nonempty, closed, bounded, and convex part of X. Consider Λ the operator given by (3.32). It follows from arguments similar to those used in the proof of (3.42) that $\Lambda\eta \in K$ for all $\eta \in X$ and, therefore, considering the restriction of Λ to K, we may assume that $\Lambda : K \to K$. Recall also that the operator Λ is weakly sequentially continuous (see Lemma 3.11) and X is separable. Therefore, we can use Theorem 1.43 to deduce that there exists an element $\eta^* \in K$ such that $\Lambda\eta^* = \eta^*$ and, moreover, η^* is a solution of the quasivariational inequality (3.29).

(2) Assume now that (3.38) also holds and let $u_1,\, u_2 \in X$ denote two solutions of (3.29); then the inequalities below hold:

$$\begin{aligned} a(u_1,\, v - u_1) + j(u_1,\, v) - j(u_1,\, u_1) &\ge (f,\, v - u_1)_X \quad \forall\, v \in X, \\ a(u_2,\, v - u_2) + j(u_2,\, v) - j(u_2,\, u_2) &\ge (f,\, v - u_2)_X \quad \forall\, v \in X. \end{aligned}$$

We take $v = u_2$ in the first inequality, $v = u_1$ in the second inequality, and add the resulting inequalities to obtain

$$a(u_1 - u_2,\, u_1 - u_2) \le j(u_1,\, u_2) - j(u_1,\, u_1) + j(u_2,\, u_1) - j(u_2,\, u_2).$$

Using now (3.2)(b), we deduce that

$$m\, \|u_1 - u_2\|_X^2 \le j(u_1,\, u_2) - j(u_1,\, u_1) + j(u_2,\, u_1) - j(u_2,\, u_2) \tag{3.47}$$

and, if $u_1 \neq u_2$, the previous inequality contradicts assumption (3.38). We conclude that $u_1 = u_2$, i.e., the quasivariational inequality (3.29) has a unique solution.

(3) We assume that (3.39) holds and note that in this case (3.38) holds, too. Therefore, by (2) we obtain that for each $f \in X$ there exists a unique solution $u = u_f$ of (3.29). Let now $f_i \in X$ and denote $u_i = u_{f_i}$, $i = 1, 2$. Using an argument similar to that in the proof of (3.47), we obtain

$$\begin{aligned} m\,\|u_1 - u_2\|_X^2 \leq j(u_1,\, u_2) - j(u_1,\, u_1) + j(u_2,\, u_1) \\ - j(u_2,\, u_2) + (f_1 - f_2, u_1 - u_2)_X \end{aligned}$$

and, using (3.39), we deduce that

$$(m - \alpha)\,\|u_1 - u_2\|_X^2 \leq (f_1 - f_2,\, u_1 - u_2)_X.$$

Since $m > \alpha$, we obtain

$$\|u_1 - u_2\|_X \leq \frac{1}{m - \alpha}\,\|f_1 - f_2\|_X,$$

which concludes the proof. □

Theorems 3.7 and 3.10 provide existence and uniqueness results in the study of the elliptic quasivariational inequality (3.29). However, besides the fact that the arguments used in their proofs are different, note that the statements of these theorems are different as well, since the assumptions on the functional j in the two theorems are different. We shall use Theorems 3.7 and 3.10 in Section 9.2 in the study of static antiplane frictional contact problems with elastic materials.

3.4 Time-dependent Elliptic Variational and Quasivariational Inequalities

The elliptic variational and quasivariational inequalities studied in this section are *time-dependent,* i.e., both the data and the solution depend on the time variable, which plays the role of a parameter. We derive existence, uniqueness, and regularity results that will be used later in this book in the study of evolutionary variational and quasivariational inequalities.

We start with the study of time-dependent variational inequalities. Let $T > 0$ and denote by $[0, T]$ the time interval of interest. Assume that $a : X \times X \to \mathbb{R}$, $j : X \to (-\infty, \infty]$ and, moreover,

$$f \in C([0,T]; X). \tag{3.48}$$

We consider the problem of finding a function $u : [0, T] \to X$ such that

$$\begin{aligned} &a(u(t), v - u(t)) + j(v) - j(u(t)) \\ &\quad \geq (f(t), v - u(t))_X \quad \forall v \in X,\ t \in [0, T]. \end{aligned} \tag{3.49}$$

We have the following existence and uniqueness result.

Theorem 3.12. *Let X be a Hilbert space and assume that* (3.2), (3.3), *and* (3.48) *hold. Then the time-dependent variational inequality* (3.49) *has a unique solution $u \in C([0, T]; X)$. Moreover, if $f \in W^{1,p}(0, T; X)$ for some $p \in [1, \infty]$, then $u \in W^{1,p}(0, T; X)$.*

Proof. For each $t \in [0, T]$, it follows from Theorem 3.1 that there exists a unique element $u(t) \in X$ that solves (3.49). To prove the continuity of the solution with respect to the time variable, we consider t_1, $t_2 \in [0, T]$. We write (3.49) for $t = t_1$ and $v = u(t_2)$, then for $t = t_2$ and $v = u(t_1)$; by adding the resulting inequalities, we obtain

$$a(u(t_1) - u(t_2), u(t_1) - u(t_2)) \leq (f(t_1) - f(t_2), u(t_1) - u(t_2))_X.$$

Then we use assumption (3.2)(b) to find that

$$m\, \|u(t_1) - u(t_2)\|_X \leq \|f(t_1) - f(t_2)\|_X. \tag{3.50}$$

Inequality (3.50) combined with the regularity (3.48) of f shows that $u \in C([0, T]; X)$, which concludes the existence part of the theorem. The uniqueness part follows from the unique solvability of (3.49) at each $t \in [0, T]$, guaranteed by Theorem 3.1.

Assume now that $f \in W^{1,p}(0, T; X)$ for some $p \in [1, \infty]$. Then $f : [0, T] \to X$ is an absolutely continuous function and inequality (3.50) shows that $u : [0, T] \to X$ is an absolutely continuous function as well, and it satisfies

$$\|\dot{u}(t)\|_X \leq \frac{1}{m}\, \|\dot{f}(t)\|_X \quad \text{a.e. } t \in (0, T). \tag{3.51}$$

Since $\dot{f} \in L^p(0, T; X)$, it follows from (3.51) that $\dot{u} \in L^p(0, T; X)$, which concludes the proof. □

Theorem 3.12 will be used in Sections 4.1 and 6.1 to obtain existence and uniqueness results in the study of evolutionary variational inequalities with viscosity and Volterra-type elliptic variational inequalities, respectively.

We now introduce a more general class of time-dependent variational inequalities and, to this end, we assume in what follows that $(Y, \|\cdot\|_Y)$ is a normed space and $\eta : [0, T] \to Y$ is a continuous function, i.e.,

$$\eta \in C([0, T]; Y). \tag{3.52}$$

We consider the problem of finding a function $u : [0,T] \to X$ such that

$$\begin{aligned} &a(u(t), v - u(t)) + j(\eta(t), v) - j(\eta(t), u(t)) \\ &\quad \geq (f(t), v - u(t))_X \quad \forall\, v \in X,\ t \in [0,T]. \end{aligned} \tag{3.53}$$

The new feature of problem (3.53) lies in the fact that j depends now on a given time-dependent function η that takes values in the space Y, which may be different from X. We assume that

$$\left.\begin{aligned} &j : Y \times X \to \mathbb{R} \text{ and} \\ &\text{(a) for all } \eta \in Y,\ j(\eta, \cdot) \text{ is convex and l.s.c. on } X. \\ &\text{(b) there exists } \alpha \geq 0 \text{ such that} \\ &\quad j(\eta_1, v_2) - j(\eta_1, v_1) + j(\eta_2, v_1) - j(\eta_2, v_2) \\ &\qquad \leq \alpha\, \|\eta_1 - \eta_2\|_Y \|v_1 - v_2\|_X \quad \forall\, \eta_1,\, \eta_2 \in Y,\ v_1,\, v_2 \in X. \end{aligned}\right\} \tag{3.54}$$

Note that assumption (3.54) is similar to assumption (3.30) and reduces to (3.30) when $Y = X$ and $\|\cdot\|_Y = \|\cdot\|_X$. Introducing a new normed space Y, different from X, is needed in Theorem 4.9 below, which will be used in Section 10.4 of the book.

The unique solvability of problem (3.53) is provided by the following result.

Theorem 3.13. *Let X be a Hilbert space, let Y be a normed space, and assume that* (3.2), (3.48), (3.52), *and* (3.54) *hold. Then there exists a unique solution u to the inequality* (3.53). *Moreover, the solution satisfies $u \in C([0,T]; X)$.*

Proof. The proof is based on arguments similar to those presented in the proof of Theorem 3.12 and for this reason we skip the details. We restrict ourselves to note that the existence and uniqueness part of the theorem follows from the unique solvability of (3.53) at each $t \in [0,T]$, guaranteed by Theorem 3.1. Consider now t_1, $t_2 \in [0,T]$; it follows from (3.53) that

$$\begin{aligned} &a(u(t_1) - u(t_2), u(t_1) - u(t_2)) \\ &\quad \leq j(\eta(t_1), u(t_2)) - j(\eta(t_1), u(t_1)) + j(\eta(t_2), u(t_1)) - j(\eta(t_2), u(t_2)) \\ &\qquad + (f(t_1) - f(t_2), u(t_1) - u(t_2))_X \end{aligned}$$

and therefore assumptions (3.2)(b) and (3.54) yield

$$m\, \|u(t_1) - u(t_2)\|_X \leq \alpha\, \|\eta(t_1) - \eta(t_2)\|_Y + \|f(t_1) - f(t_2)\|_X. \tag{3.55}$$

Inequality (3.55) combined with the regularity (3.48) and (3.52) shows that $u \in C([0,T]; X)$, which concludes the proof. □

We now consider the case of time-dependent quasivariational inequalities. Let $T > 0$ and denote again by $[0,T]$ the time interval of interest. We consider the problem of finding a function $u : [0, T] \to X$ such that

$$\begin{aligned} a(u(t), v - u(t)) + j(u(t), v) - j(u(t), u(t)). \\ \geq (f(t), v - u(t))_X \quad \forall v \in X,\ t \in [0, T] \end{aligned} \tag{3.56}$$

We have the following existence and uniqueness result.

Theorem 3.14. *Let X be a Hilbert space and assume that* (3.2), (3.30), *and* (3.48) *hold. Then, if $m > \alpha$, there exists a unique solution u to the inequality* (3.56). *Moreover, the solution satisfies $u \in C([0, T]; X)$.*

Proof. A first proof of the theorem can be obtained as follows: since $m > \alpha$, it follows from Theorem 3.7 that for each $t \in [0, T]$ there exists a unique element $u(t) \in X$ that solves (3.56). To prove the continuity of the solution with respect to the time variable consider $t_1,\ t_2 \in [0, T]$. Arguments similar to those used to derive (3.35) yield

$$\|u(t_1) - u(t_2)\|_X \leq \frac{1}{m - \alpha}\, \|f(t_1) - f(t_2)\|_X, \tag{3.57}$$

which implies the continuity of the solution and concludes the existence part of the theorem. The uniqueness follows from the unique solvability of (3.53) at each $t \in [0, T]$, guaranteed by Theorem 3.7.

A second proof of the theorem is based on a fixed point argument. It follows from Theorem 3.13 that for each element $\eta \in C([0, T]; X)$, there exists a unique solution to problem (3.56), denoted u_η, and it satisfies $u_\eta \in C([0, T]; X)$. Let $\Lambda : C([0, T]; X) \to C([0, T]; X)$ be the operator given by

$$\Lambda\eta = u_\eta \quad \forall \eta \in C([0, T]; X). \tag{3.58}$$

Arguments similar to those used in the proof of (3.55) show that

$$m\, \|u_1(t) - u_2(t)\|_X \leq \alpha\, \|\eta_1(t) - \eta_2(t)\|_X \quad \forall t \in [0, T],$$

where u_1 and u_2 are the solutions of (3.56) associated with the functions η_1 and η_2, both in $C([0, T]; X)$. If $m > \alpha$, it results from the previous inequality that the operator Λ is a contraction on the Banach space $C([0, T]; X)$, which concludes the existence part of the theorem. The uniqueness part follows from the uniqueness of the fixed point of the operator (3.58). □

Chapter 4
Evolutionary Variational Inequalities with Viscosity

The variational inequalities studied in this chapter are evolutionary since they involve the time derivative of the solution and an initial condition. Their main feature lies in the fact that they are governed by two bilinear forms, one of which involves only the time derivative of the solution, consequently, using the terminology arising in the mechanics of continua, we call then *evolutionary variational inequality with viscosity.* We start with a basic existence and uniqueness result then we prove regularity and convergence results. We also consider evolutionary quasivariational inequalities and history-dependent evolutionary variational inequalities with viscosity for which we prove existence and uniqueness results. The results presented in this chapter have interest in and of themselves, as they may be applied directly in the study of frictional contact problems involving viscoelastic materials with short memory. Also, they are useful in the study of evolutionary variational inequalities via the regularization method, as it will be shown in the next chapter. As usual, everywhere in this chapter X denotes a real Hilbert space with inner product $(\cdot,\cdot)_X$ and norm $\|\cdot\|_X$ and, moreover, $[0,T]$ denotes the time interval of interest, where $T>0$.

4.1 A Basic Existence and Uniqueness Result

Let a and b denote two bilinear forms on X, $j : X \to (-\infty, \infty]$, $f : [0,T] \to X$ and let u_0 be an initial data. In this section, we are interested in sufficient conditions for the existence and uniqueness of the solution of the following problem: find $u : [0,T] \to X$ such that

$$\begin{aligned} &a(u(t), v-\dot{u}(t)) + b(\dot{u}(t), v-\dot{u}(t)) + j(v) - j(\dot{u}(t)) \\ &\quad \geq (f(t), v-\dot{u}(t))_X \quad \forall\, v\in X,\ t\in[0,T], \end{aligned} \tag{4.1}$$

$$u(0) = u_0. \tag{4.2}$$

Inequality (4.1) represents an *evolutionary variational inequality* since it involves the time derivative of the unknown function u. As a consequence, an initial condition, (4.2), is needed. The term $b(\dot{u}(t), v - \dot{u}(t))$ in (4.1) is called the *viscosity term*; the origin of this terminology arises from mechanics and will be explained in Part IV of this book. Notice that this term simplifies considerably the analysis of the problem, since it produces a regularization effect; its presence allows us to prove the unique solvability of the problem (4.1)–(4.2) by using only arguments of time-dependent elliptic variational inequalities and fixed point.

In the study of the Cauchy problem (4.1)–(4.2), we assume that:

$$\left.\begin{array}{l} a : X \times X \to \mathbb{R} \text{ is a bilinear form and} \\ \quad \text{there exists } M > 0 \text{ such that} \\ \qquad |a(u,v)| \le M\|u\|_X\|v\|_X \quad \forall\, u,\, v \in X. \end{array}\right\} \tag{4.3}$$

$$\left.\begin{array}{l} b : X \times X \to \mathbb{R} \text{ is a bilinear symmetric form and} \\ \text{(a) there exists } M' > 0 \text{ such that} \\ \qquad |b(u,v)| \le M'\|u\|_X\|v\|_X \quad \forall\, u,\, v \in X. \\ \text{(b) there exists } m' > 0 \text{ such that} \\ \qquad b(v,v) \ge m'\|v\|_X^2 \quad \forall\, v \in X. \end{array}\right\} \tag{4.4}$$

$$j : X \to (-\infty, \infty] \text{ is a proper convex l.s.c. function.} \tag{4.5}$$

$$f \in C([0,T];X). \tag{4.6}$$

$$u_0 \in X. \tag{4.7}$$

Note that assumption (4.3) shows that the bilinear form a is continuous, whereas assumption (4.4) shows that the bilinear form b is symmetric, continuous, and X-elliptic. The X-ellipticity of the bilinear form a is not needed, nor its symmetry, and this is due to the presence of the bilinear form b in (4.1), which plays a crucial role in the study of problem (4.1)–(4.2).

The main result of this section is the following.

Theorem 4.1. *Let X be a Hilbert space and assume that* (4.3)–(4.7) *hold. Then there exists a unique solution u to problem* (4.1)–(4.2). *Moreover, the solution satisfies $u \in C^1([0,T];X)$.*

The proof of the Theorem 4.1 will be carried out in three steps. The main idea is to start by solving an intermediate problem for the velocity function $\dot{u}$ and then to use a fixed point argument. We assume in what follows that (4.3)–(4.7) hold.

In the first step, let $\eta \in C([0,T];X)$ be given and consider the problem of finding $w_\eta : [0,T] \to X$ such that

$$\begin{aligned} &a(\eta(t), v - w_\eta(t)) + b(w_\eta(t), v - w_\eta(t)) + j(v) - j(w_\eta(t)) \\ &\quad \geq (f(t), v - w_\eta(t))_X \quad \forall\, v \in X,\ t \in [0,T]. \end{aligned} \tag{4.8}$$

The unique solvability of this problem is given by the following result.

Lemma 4.2. *There exists a unique solution w_η to the inequality* (4.8)*. Moreover, the solution satisfies $w_\eta \in C([0,T];X)$.*

Proof. We use (4.3) and Riesz's representation theorem to define the function $f_\eta : [0,T] \to X$ by

$$(f_\eta(t), v)_X = (f(t), v)_X - a(\eta(t), v) \quad \forall\, v \in X,\ t \in [0,T]. \tag{4.9}$$

From assumption (4.3) on the form a, combined with the regularity $f \in C([0,T];X)$ and $\eta \in C([0,T];X)$, it follows that $f_\eta \in C([0,T];X)$. Therefore, assumptions (4.4) and (4.5) allow us to apply Theorem 3.12 to deduce that there exists a unique function $w_\eta \in C([0,T];X)$ that solves the time-dependent variational inequality

$$\begin{aligned} &b(w_\eta(t), v - w_\eta(t)) + j(v) - j(w_\eta(t)) \\ &\quad \geq (f_\eta(t), v - w_\eta(t))_X \quad \forall\, v \in X,\ t \in [0,T]. \end{aligned} \tag{4.10}$$

We combine (4.9) and (4.10) to see that the element $w_\eta \in C([0,T];X)$ is the unique solution to the variational inequality (4.8). □

Next, we consider the operator $\Lambda : C([0,T];X) \to C([0,T];X)$ defined by

$$\Lambda\eta(t) = \int_0^t w_\eta(s)ds + u_0 \quad \forall\, \eta \in C([0,T];X),\ t \in [0,T]. \tag{4.11}$$

The second step is given by the following result.

Lemma 4.3. *The operator Λ has a unique fixed point $\eta^* \in C([0,T];X)$.*

Proof. Let $\eta_1, \eta_2 \in C([0,T];X)$ and denote by w_i the solution of (4.8) for $\eta = \eta_i$, i.e., $w_i = w_{\eta_i}$, $i = 1,2$. From the definition (4.11), we deduce the inequality

$$\|\Lambda\,\eta_1(t) - \Lambda\,\eta_2(t)\|_X \leq \int_0^t \|w_1(s) - w_2(s)\|_X\, ds \quad \forall\, t \in [0,T]. \tag{4.12}$$

On the other hand, using (4.8) for $\eta = \eta_i$, $i = 1,2$, we obtain the inequality

$$b(w_1(s) - w_2(s), w_1(s) - w_2(s)) \leq a(\eta_1(s) - \eta_2(s), w_2(s) - w_1(s)) \quad \forall\, s \in [0,T]$$

and, using the assumptions (4.3) and (4.4), we find

$$\|w_1(s) - w_2(s)\|_X \leq \frac{M}{m'}\, \|\eta_1(s) - \eta_2(s)\|_X \quad \forall\, s \in [0, T]. \tag{4.13}$$

We combine inequalities (4.12) and (4.13) to obtain

$$\|\Lambda\,\eta_1(t) - \Lambda\,\eta_2(t)\|_X \leq c \int_0^t \|\eta_1(s) - \eta_2(s)\|_X \, ds \quad \forall\, t \in [0, T] \tag{4.14}$$

with $c = \dfrac{M}{m'}$. Lemma 4.3 is now a direct consequence of inequality (4.14) and Lemma 1.42 (page 16). □

We now have all the ingredients to provide the proof of Theorem 4.1.

Proof. Existence. Let $\eta^* \in C([0,T]; X)$ be the fixed point of Λ and let $u = \eta^*$. Since $\eta^* = \Lambda\eta^*$, it follows from (4.11) that $u(0) = u_0$ and $\dot{u} = w_{\eta^*}$; therefore, we conclude by (4.8) that u is a solution to problem (4.1)–(4.2). Moreover, since $\dot{u} = w_{\eta^*}$ and $w_{\eta^*} \in C([0,T]; X)$, we deduce that u has the regularity $u \in C^1([0,T]; X)$.

Uniqueness. The uniqueness of the solution follows from the uniqueness of the fixed point of the operator Λ defined by (4.11), by using arguments similar to those used in the proof of the uniqueness part of Theorem 3.7. It also can be obtained directly from (4.1)–(4.2), as follows: assume that u_1, $u_2 \in C^1([0,T]; X)$ are two solutions of (4.1)–(4.2) and let $t \in [0,T]$. We have

$$\begin{aligned}
&a(u_1(t), v - \dot{u}_1(t)) + b(\dot{u}_1(t), v - \dot{u}_1(t)) + j(v) - j(\dot{u}_1(t)) \\
&\qquad \geq (f(t), v - \dot{u}_1(t))_X \quad \forall\, v \in X, \\
&a(u_2(t), v - \dot{u}_2(t)) + b(\dot{u}_2(t), v - \dot{u}_2(t)) + j(v) - j(\dot{u}_2(t)) \\
&\qquad \geq (f(t), v - \dot{u}_2(t))_X \quad \forall\, v \in X.
\end{aligned}$$

We take $v = \dot{u}_2(t)$ in the first inequality, $v = \dot{u}_1(t)$ in the second one, and add the results to obtain

$$b(\dot{u}_1(t) - \dot{u}_2(t), \dot{u}_1(t) - \dot{u}_2(t)) \leq a(u_1(t) - u_2(t), \dot{u}_2(t) - \dot{u}_1(t)). \tag{4.15}$$

We use now assumptions (4.3) and (4.4) to find

$$\|\dot{u}_1(t) - \dot{u}_2(t)\|_X \leq \frac{M}{m'}\, \|u_1(t) - u_2(t)\|_X. \tag{4.16}$$

Moreover, since $u_i(0) = u_0$ it follows that

$$u_i(t) = \int_0^t \dot{u}_i(s)\, ds + u_0 \quad \forall\, i = 1, 2, \tag{4.17}$$

and, therefore,

$$\|u_1(t) - u_2(t)\|_X \le \int_0^t \|\dot{u}_1(s) - \dot{u}_2(s)\|_X \, ds. \tag{4.18}$$

We combine now inequalities (4.16) and (4.18) and use Gronwall's inequality (Lemma 1.44, page 17) to deduce the equality $\dot{u}_1(s) = \dot{u}_2(s)$, for all $s \in [0,T]$; using then (4.17) we obtain that $u_1(t) = u_2(t)$, which concludes the proof. □

Applying Theorem 4.1 to the case when j is the indicator function of a nonempty, convex, and closed subset $K \subset X$ leads to the following result.

Corollary 4.4. *Let X be a Hilbert space, let $K \subset X$ be a nonempty, convex, and closed subset, and assume that (4.3), (4.4), (4.6), (4.7) hold. Then there exists a unique function $u \in C^1([0,T];X)$ such that $u(0) = u_0$ and*

$$\begin{aligned} \dot{u}(t) \in K, \quad & a(u(t), v - \dot{u}(t)) + b(\dot{u}(t), v - \dot{u}(t)) \\ & \ge (f(t), v - \dot{u}(t))_X \quad \forall\, v \in K,\ t \in [0,T]. \end{aligned}$$

We end this section with a regularity result.

Theorem 4.5. *Under the conditions stated in Theorem 4.1, if there exists $p \in [1,\infty]$ such that $f \in W^{1,p}(0,T;X)$, then $u \in W^{2,p}(0,T;X)$.*

Proof. For any $t_1,\ t_2 \in [0,T]$, we use (4.1) and the arguments used in the proof of (4.15) to obtain

$$\begin{aligned} & b(\dot{u}(t_1) - \dot{u}(t_2),\, \dot{u}(t_1) - \dot{u}(t_2)) \\ & \quad \le a(u(t_1) - u(t_2), \dot{u}(t_2) - \dot{u}(t_1)) + (f(t_1) - f(t_2), \dot{u}(t_1)) - \dot{u}(t_2))_X. \end{aligned}$$

We employ now assumptions (4.3) and (4.4) to find

$$\|\dot{u}(t_1) - \dot{u}(t_2)\|_X \le c\,(\|u(t_1) - u(t_2)\|_X + \|f(t_1) - f(t_2)\|_X) \tag{4.19}$$

where $c = \max\{\frac{M}{m'}, \frac{1}{m'}\}$. This inequality combined with the regularity $u \in C^1([0,T];X)$, $f \in W^{1,p}(0,T;X)$ shows that $\dot{u} : [0,T] \to X$ is an absolutely continuous function and, moreover,

$$\|\ddot{u}(t)\|_X \le c\,(\|\dot{u}(t)\|_X + \|\dot{f}(t)\|_X) \ \text{ a.e. } t \in (0,T).$$

We conclude that $\ddot{u} \in L^p(0,T;X)$ and therefore $u \in W^{2,p}(0,T;X)$. □

4.2 A Convergence Result

In this section, we study the dependence of the solution of the variational inequality (4.1)–(4.2) on the function j. To this end, we assume in what follows that (4.3)–(4.7) hold and we denote by $u \in C^1([0,T];X)$ the solution of (4.1)–(4.2) provided by Theorem 4.1.

For every $\rho > 0$, we also consider a perturbation j_ρ of j that satisfies (4.5) and we consider the problem of finding a function $u_\rho : [0,T] \to X$ such that

$$a(u_\rho(t), v - \dot{u}_\rho(t)) + b(\dot{u}_\rho(t), v - \dot{u}_\rho(t)) + j_\rho(v) - j_\rho(\dot{u}_\rho(t)) \geq (f(t), v - \dot{u}_\rho(t))_X \quad \forall v \in X,\ t \in [0,T], \tag{4.20}$$

$$u_\rho(0) = u_0. \tag{4.21}$$

We deduce from Theorem 4.1 that problem (4.20)–(4.21) has a unique solution $u_\rho \in C^1([0,T];X)$, for each $\rho > 0$.

Our main result in this section is the following.

Theorem 4.6. *Under the assumption* (3.17), *the solution* u_ρ *of problem* (4.20)–(4.21) *converges to the solution* u *of problem* (4.1)–(4.2), *i.e.,*

$$\|u_\rho - u\|_{C^1([0,T];X)} \to 0 \quad \text{as } \rho \to 0. \tag{4.22}$$

Proof. Let $\rho > 0$ and let $t \in [0,T]$. We take $v = \dot{u}_\rho(t)$ in (4.1), $v = \dot{u}(t)$ in (4.20) and add the resulting inequalities to obtain

$$\begin{aligned} b(\dot{u}_\rho(t) - \dot{u}(t), \dot{u}_\rho(t) - \dot{u}(t)) &\leq a(u_\rho(t) - u(t), \dot{u}(t) - \dot{u}_\rho(t)) \\ &\quad + j(\dot{u}_\rho(t)) - j(\dot{u}(t)) + j_\rho(\dot{u}(t)) - j_\rho(\dot{u}_\rho(t)) \\ &\leq a(u_\rho(t) - u(t), \dot{u}(t) - \dot{u}_\rho(t)) \\ &\quad + |j(\dot{u}_\rho(t)) - j_\rho(\dot{u}_\rho(t))| + |j_\rho(\dot{u}(t)) - j(\dot{u}(t))|. \end{aligned}$$

We use assumptions (4.3), (4.4) and (3.17) to find that

$$m'\|\dot{u}_\rho(t) - \dot{u}(t)\|_X^2 \leq M\|u_\rho(t) - u(t)\|_X \|\dot{u}_\rho(t) - \dot{u}(t)\|_X + 2\,F(\rho)$$

and, using the inequality

$$M\|u_\rho(t) - u(t)\|_X \|\dot{u}_\rho(t) - \dot{u}(t)\|_X \leq \frac{M^2}{2m'}\,\|u_\rho(t) - u(t)\|_X^2 + \frac{m'}{2}\,\|\dot{u}_\rho(t) - \dot{u}(t)\|_X^2,$$

we find that

$$\|\dot{u}_\rho(t) - \dot{u}(t)\|_X^2 \leq \frac{4}{m'}\,F(\rho) + \frac{M^2}{m'^2}\,\|u_\rho(t) - u(t)\|_X^2. \tag{4.23}$$

It follows from (4.21) and (4.2) that

$$\|u_\rho(t) - u(t)\|_X \leq \int_0^t \|\dot{u}_\rho(s) - \dot{u}(s)\|_X\, ds, \tag{4.24}$$

which implies that

$$\|u_\rho(t) - u(t)\|_X^2 \le T \int_0^t \|\dot{u}_\rho(s) - \dot{u}(s)\|_X^2 \, ds.$$

Substituting this inequality in (4.23) leads to

$$\|\dot{u}_\rho(t) - \dot{u}(t)\|_X^2 \le \frac{4}{m'} \, F(\rho) + \frac{M^2 T}{m'^2} \int_0^t \|\dot{u}_\rho(s) - \dot{u}(s)\|_X^2 \, ds.$$

We use now Gronwall's inequality (page 17) to see that

$$\|\dot{u}_\rho(t) - \dot{u}(t)\|_X^2 \le \frac{4}{m'} \, e^{\frac{M^2 T^2}{m'^2}} \, F(\rho). \tag{4.25}$$

Finally, we combine inequalities (4.24) and (4.25) and use assumption (3.17) to obtain (4.22). □

Theorem 4.6 shows that, under condition (3.17), the solution u of the evolutionary variational inequality (4.1)–(4.2) can be approached by the solution of the evolutionary variational inequality (4.20)–(4.21), in the sense of (4.22). Assume now that j_ρ is a Gâteaux differentiable function and denote by $B : X \to X$ the operator given by

$$(Bu, v)_X = b(u, v) \quad \forall\, u, v \in X. \tag{4.26}$$

We use the notation (3.27), (3.28), and (4.26) to see that problem (4.20)–(4.21) is equivalent to the Cauchy problem

$$Au_\rho(t) + B\dot{u}_\rho(t) + \nabla j_\rho(u_\rho(t)) = f(t) \quad \forall\, t \in [0, T], \tag{4.27}$$

$$u_\rho(0) = u_0. \tag{4.28}$$

We conclude that solving the variational inequality (4.1)–(4.2) leads to the solution of the Cauchy problem (4.27)–(4.28) for a "small" parameter ρ.

4.3 Evolutionary Quasivariational Inequalities with Viscosity

In this section, we consider evolutionary quasivariational inequalities with viscosity, i.e., we allow the functional j to depend explicitly on the solution u or on its derivative $\dot{u}$. Thus, we assume that $j : X \times X \to \mathbb{R}$ and we consider the following problem: find $u : [0, T] \to X$ such that

$$a(u(t), v - \dot{u}(t)) + b(\dot{u}(t), v - \dot{u}(t)) + j(u(t), v) - j(u(t), \dot{u}(t)) \\ \ge (f(t), v - \dot{u}(t))_X \quad \forall\, v \in X, \ t \in [0, T], \tag{4.29}$$

$$u(0) = u_0. \tag{4.30}$$

We also consider the problem of finding $u : [0,T] \to X$ such that

$$a(u(t), v - \dot{u}(t)) + b(\dot{u}(t), v - \dot{u}(t)) + j(\dot{u}(t), v) - j(\dot{u}(t), \dot{u}(t)) \\ \geq (f(t), v - \dot{u}(t))_X \quad \forall v \in X,\ t \in [0,T], \tag{4.31}$$

$$u(0) = u_0. \tag{4.32}$$

In the study of these evolutionary quasivariational inequalities, we have the following existence and uniqueness result.

Theorem 4.7. *Let X be a Hilbert space and assume that* (4.3), (4.4), (4.6), (4.7), *and* (3.30) *hold. Then:*

(1) There exists a unique solution $u \in C^1([0,T];X)$ to problem (4.29)–(4.30).

(2) If $m' > \alpha$, there exists a unique solution $u \in C^1([0,T];X)$ to problem (4.31)–(4.32).

Proof. (1) We use arguments similar to those used in the proof of Theorem 4.1, based on the unique solvability of time-dependent elliptic variational inequalities and the Banach fixed point theorem. Since the modifications are straightforward, we omit the details. The steps of the proof are the following.

(i) Let $\eta \in C([0,T];X)$ be given. We use Theorem 3.13 with $Y = X$ to see that there exists a unique element $w_\eta \in C([0,T];X)$ such that

$$a(\eta(t), v - w_\eta(t)) + b(w_\eta(t), v - w_\eta(t)) + j(\eta(t), v) - j(\eta(t), w_\eta(t)) \\ \geq (f(t), v - w_\eta(t))_X \quad \forall v \in X,\ t \in [0,T]. \tag{4.33}$$

(ii) Next, we consider the operator $\Lambda : C([0,T];X) \to C([0,T];X)$ defined by (4.11) and we prove that it has a unique fixed point $\eta^* \in C([0,T];X)$. Indeed, let $\eta_1,\ \eta_2 \in C([0,T];X)$ and denote by w_i the solutions of (4.33) for $\eta = \eta_i$, i.e., $w_i = w_{\eta_i}$, $i = 1,2$. It follows that

$$b(w_1(s) - w_2(s), w_1(s) - w_2(s)) \\ \leq a(\eta_1(s) - \eta_2(s), w_2(s) - w_1(s)) + j(\eta_1(s), w_2(s)) - j(\eta_1(s), w_1(s)) \\ + j(\eta_2(s), w_1(s)) - j(\eta_2(s), w_2(s)) \quad \forall s \in [0,T]$$

and, using the assumptions (4.3), (4.4), and (3.30), we obtain

$$\|w_1(s) - w_2(s)\|_X \leq \frac{M + \alpha}{m'}\, \|\eta_1(s) - \eta_2(s)\|_X \quad \forall s \in [0,T]. \tag{4.34}$$

We combine (4.12) and (4.34) to find (4.14) with $c = \dfrac{M+\alpha}{m'}$ and then use Lemma 1.42.

(iii) Let $\eta^* \in C([0,T];X)$ be the fixed point of Λ and let $u = \eta^*$. Then it follows that u is the unique solution of problem (4.29)–(4.30) and, moreover, it satisfies $u \in C^1([0,T];X)$.

(2) This time we use arguments on time-dependent elliptic quasivariational inequalities. The steps of the proof are the following.

(i) Let $\eta \in C([0,T];X)$ be given. We use Theorem 3.14 to see that, if $m' > \alpha$, there exists a unique element $w_\eta \in C([0,T];X)$ such that

$$\begin{aligned} &a(\eta(t), v - w_\eta(t)) + b(w_\eta(t), v - w_\eta(t)) + j(w_\eta(t), v) \\ &\quad - j(w_\eta(t), w_\eta(t)) \geq (f(t), v - w_\eta(t))_X \quad \forall\, v \in X,\ t \in [0,T]. \end{aligned} \tag{4.35}$$

(ii) Next, we consider the operator $\Lambda : C([0,T];X) \to C([0,T];X)$ defined by (4.11) and we prove again that it has a unique fixed point. Indeed, if w_i represents the solution of (4.35) for $\eta = \eta_i \in C([0,T];X)$, $i = 1,2$, it follows that

$$\begin{aligned} &b(w_1(s) - w_2(s), w_1(s) - w_2(s)) \\ &\quad \leq a(\eta_1(s) - \eta_2(s), w_2(s) - w_1(s)) + j(w_1(s), w_2(s)) - j(w_1(s), w_1(s)) \\ &\qquad + j(w_2(s), w_1(s)) - j(w_2(s), w_2(s)) \quad \forall s \in [0,T]. \end{aligned}$$

Now, by using assumptions (4.3), (4.4), (3.30) and the additional assumption $m' > \alpha$, we find that

$$\|w_1(s) - w_2(s)\|_X \leq \frac{M}{m' - \alpha}\, \|\eta_1(s) - \eta_2(s)\|_X \quad \forall s \in [0,T].$$

We combine this inequality with (4.12) to find (4.14) with $c = \dfrac{M}{m' - \alpha}$ and then use again Lemma 1.42.

(iii) Let $\eta^* \in C([0,T];X)$ be the fixed point of Λ and let $u = \eta^*$. Then it follows that u is the unique solution of problem (4.31)–(4.32) and, moreover, it satisfies $u \in C^1([0,T];X)$. □

It follows from Theorem 4.7 that the unique solvability of the evolutionary quasivariational inequalities (4.29)–(4.30) and (4.31)–(4.32), respectively, is proved under common assumptions that include (4.3), (4.4), (4.6), (4.7), and (3.30); also, note that the regularity $u \in C^1([0,T];X)$ of the solution in both of the two problems is the same. Therefore, we conclude that these variational inequalities have similar features. However, we note that there exists a difference between these problems, and it arises from the fact that in solving problem (4.29)–(4.30), we do not need any additional assumptions, whereas in solving problem (4.31)–(4.32) we need to assume that $m' > \alpha$, which is a *smallness assumption* on the coefficient α. Examples of boundary value problems that lead to evolutionary quasivariational inequalities of the form (4.29)–(4.30) and (4.31)–(4.32) will be presented in Section 10.3 of the book.

We end this section with a regularity result, similar to that presented in Theorem 4.5.

Theorem 4.8. *Under the conditions stated in Theorem 4.7, denote by u the solution of problem (4.29)–(4.30) or (4.31)–(4.32) and assume, moreover, that $f \in W^{1,p}(0,T;X)$ for some $p \in [1,\infty]$. Then $u \in W^{2,p}(0,T;X)$.*

Proof. The proof is similar to that of Theorem 4.5 and is based on the estimate (4.19); it is easy to see that this estimate holds in the study of problem (4.29)–(4.30) with $c = \max\{\frac{M+\alpha}{m'}, \frac{1}{m'}\}$ and in the study of problem (4.31)–(4.32) with $c = \max\{\frac{M}{m'-\alpha}, \frac{1}{m'-\alpha}\}$. Recall that in this last case, as stated in Theorem 4.7(2), we assume that $m' > \alpha$. □

4.4 History-dependent Evolutionary Variational Inequalities with Viscosity

We turn now to the study of variational inequalities with viscosity for which the functional j depends on the integral of the solution or on the integral of its derivative. To this end, let $(Y, \|\cdot\|_Y)$ be a Hilbert space, assume that $j : Y \times X \to \mathbb{R}$, and consider an operator $S : C([0,T];X) \to C([0,T];Y)$. We are interested in providing conditions that guarantee the unique solvability of the following problem: find $u : [0,T] \to X$ such that

$$\begin{aligned} a(u(t), v - \dot{u}(t)) + b(\dot{u}(t), v - \dot{u}(t)) + j(Su(t), v) - j(Su(t), \dot{u}(t)) \\ \geq (f(t), v - \dot{u}(t))_X \quad \forall\, v \in X,\ t \in [0,T], \end{aligned} \tag{4.36}$$

$$u(0) = u_0. \tag{4.37}$$

We also are interested in the unique solvability of the problem of finding $u : [0,T] \to X$ such that

$$\begin{aligned} a(u(t), v - \dot{u}(t)) + b(\dot{u}(t), v - \dot{u}(t)) + j(S\dot{u}(t), v) - j(S\dot{u}(t), \dot{u}(t)) \\ \geq (f(t), v - \dot{u}(t))_X \quad \forall\, v \in X,\ t \in [0,T], \end{aligned} \tag{4.38}$$

$$u(0) = u_0. \tag{4.39}$$

In the study of these problems, we assume that the operator S satisfies the following condition:

$$\left.\begin{array}{l} \text{there exists } L_S > 0 \text{ such that} \\ \displaystyle \|Sv_1(t) - Sv_2(t)\|_Y \leq L_S \int_0^t \|v_1(s) - v_2(s)\|_X\, ds \\ \forall\, v_1,\, v_2 \in C([0,T];X),\ t \in [0,T]. \end{array}\right\} \tag{4.40}$$

Note that this condition is satisfied by the operator $S : C([0,T];X) \to C([0,T];Y)$ given by

$$Sv(t) = \int_0^t Av(s)\, ds + y_0 \quad \forall\, v \in C([0,T];X),\ t \in [0,T], \tag{4.41}$$

where $A : X \to Y$ is a Lipschitz continuous operator and $y_0 \in Y$. It is also satisfied for the *Volterra operator* $S : C([0,T];X) \to C([0,T];Y)$ given by

$$Sv(t) = \int_0^t A(t-s)\,v(s)\,ds + y_0 \quad \forall\, v \in C([0,T];X),\ t \in [0,T], \tag{4.42}$$

where now $A \in C([0,T];\mathcal{L}(X,Y))$ and, again, $y_0 \in Y$. Indeed, in the case of the operator (4.41), inequality (4.40) holds with L_S being the Lipschitz constant of the operator A, and in the case of the operator (4.42), it holds with

$$L_S = \|A\|_{C([0,T];\mathcal{L}(X,Y))} = \max_{t\in[0,T]} \|A(t)\|_{\mathcal{L}(X,Y)}.$$

Clearly, for the operators (4.41) and (4.42), the current value $Sv(t)$ at the moment $t \in [0,T]$ depends on the history of the values of v at the moments $0 \le s \le t$ and, therefore, we refer to operators of the form (4.41) or (4.42) as *history-dependent operators.* We extend this definition to all the operators $S : C([0,T];X) \to C([0,T];Y)$ satisfying condition (4.40) and, for this reason, we say that the variational inequalities (4.36)–(4.37) and (4.38)–(4.39) are *history-dependent variational inequalities.* Their new feature consists in the fact that, at each moment $t \in [0,T]$, the functional j depends on the *history of the solution* up to the moment t, $Su(t)$, or on the *history of the derivative of the solution* up to the moment t, $S\dot{u}(t)$. This feature makes these problems different from the quasivariational inequalities studied in Section 4.3, since there j was assumed to depend on the *current value of the solution*, $u(t)$, or on the *current value of its derivative*, $\dot{u}(t)$.

In the study of history-dependent evolutionary variational inequalities with viscosity, we have the following existence and uniqueness result.

Theorem 4.9. *Let X and Y be two Hilbert spaces and assume that* (4.3), (4.4), (4.6), (4.7), (3.54), *and* (4.40) *hold. Then:*

(1) There exists a unique solution $u \in C^1([0,T];X)$ to problem (4.36)–(4.37).

(2) There exists a unique solution $u \in C^1([0,T];X)$ to problem (4.38)–(4.39).

Proof. (1) We use arguments similar to those used in the proofs of Theorems 4.1 and 4.7 but with a different choice of the operator Λ. Below, c will represent various positive constants whose values may change from line to line. The steps of the proof are the following.

(i) Let $\eta = (\eta^1, \eta^2) \in C([0,T];X \times Y)$ be given. We use Theorem 3.13 to see that there exists a unique element $w_\eta \in C([0,T];X)$ such that

$$\begin{aligned} &a(\eta^1(t), v - w_\eta(t)) + b(w_\eta(t), v - w_\eta(t)) + j(\eta^2(t), v) - j(\eta^2(t), w_\eta(t)) \\ &\quad \ge (f(t), v - w_\eta(t))_X \quad \forall\, v \in X,\ t \in [0,T]. \end{aligned} \tag{4.43}$$

(ii) We consider the operator $\Lambda : C([0,T]; X \times Y) \to C([0,T]; X \times Y)$ defined by

$$\Lambda\eta(t) = \Big(\int_0^t w_\eta(s)\, ds + u_0,\ S\Big(\int_0^t w_\eta(s)\, ds + u_0 \Big) \Big)$$
$$\forall\, \eta \in C([0,T]; X \times Y),\ t \in [0,T], \tag{4.44}$$

and prove that Λ has a unique fixed point η^*. To this end, let $\eta = (\eta^1, \eta^2)$ and $\xi = (\xi^1, \xi^2)$ be two elements of $C([0,T]; X \times Y)$ and denote by w_η and w_ξ the corresponding solutions of inequality (4.43). After performing some algebraic manipulations, we find

$$\begin{aligned} & b(w_\eta(s) - w_\xi(s), w_\eta(s) - w_\xi(s)) \\ & \quad \le a(\eta^1(s) - \xi^1(s), w_\xi(s) - w_\eta(s)) + j(\eta^2(s), w_\xi(s)) - j(\eta^2(s), w_\eta(s)) \\ & \qquad + j(\xi^2(s), w_\eta(s)) - j(\xi^2(s), w_\xi(s)) \end{aligned}$$

and, using assumptions (4.3), (4.4), and (3.54), we deduce that

$$\begin{aligned} \|w_\eta(s) - w_\xi(s)\|_X \le \frac{M}{m'}\, & \|\eta^1(s) - \xi^1(s)\|_X \\ & + \frac{\alpha}{m'}\, \|\eta^2(s) - \xi^2(s)\|_Y \quad \forall\, s \in [0,T]. \end{aligned} \tag{4.45}$$

We conclude by (4.45) that

$$\|w_\eta(s) - w_\xi(s)\|_X \le c\, \|\eta(s) - \xi(s)\|_{X \times Y} \quad \forall\, s \in [0,T] \tag{4.46}$$

where $\|\cdot\|_{X \times Y}$ denotes the norm on the space $X \times Y$, given by

$$\|\eta\|_{X \times Y} = \big(\|\eta^1\|_X^2 + \|\eta^2\|_Y^2 \big)^{\frac{1}{2}} \quad \forall\, \eta = (\eta^1, \eta^2) \in X \times Y.$$

Now, (4.44), (4.46), and (4.40) yield

$$\begin{aligned} & \|\Lambda\eta(t) - \Lambda\xi(t)\|_{X \times Y} \\ & \quad \le c \int_0^t \|\eta(s) - \xi(s)\|_{X \times Y} + c \int_0^t \Big(\int_0^s \|\eta(r) - \xi(r)\|_{X \times Y} dr \Big) ds \ \ \forall\, t \in [0,T] \end{aligned} \tag{4.47}$$

and, since

$$\int_0^t \Big(\int_0^s \|\eta(r) - \xi(r)\|_{X \times Y}\, dr \Big)\, ds \le T \int_0^t \|\eta(s) - \xi(s)\|_{X \times Y}\, ds,$$

from (4.47) we obtain that

$$\|\Lambda\eta(t) - \Lambda\xi(t)\|_{X \times Y} \le c \int_0^t \|\eta(s) - \xi(s)\|_{X \times Y}\, ds. \tag{4.48}$$

Inequality (4.48) and Lemma 1.42 show that the operator Λ has a unique fixed point.

(iii) Let $\eta^* = (\eta^{1*}, \eta^{2*}) \in C([0,T]; X \times Y)$ be the fixed point of Λ and let $u \in C^1([0,T]; X)$ be the function defined by

$$u(t) = \int_0^t w_{\eta^*}(s)\, ds + u_0 \quad \forall\, t \in [0,T]. \tag{4.49}$$

It follows from (4.44) and (4.49) that

$$\eta^{1*} = u, \qquad \eta^{2*} = Su \tag{4.50}$$

and, writing (4.43) for $\eta = \eta^*$ and using (4.50), we deduce that u satisfies inequality (4.36). Also, (4.37) is a consequence of (4.49) and, since $w_{\eta^*} \in C([0,T]; X)$, we deduce that $u \in C^1([0,T]; X)$, which concludes the existence part of Theorem 4.9(1).

(iv) The uniqueness part is a consequence of the uniqueness of the fixed point of the operator (4.44). It can also be obtained directly from (4.36), (4.37), as follows: assume that $u_1,\, u_2 \in C^1([0,T]; X)$ are two functions that satisfy (4.36)–(4.37) and let $t \in [0,T]$. Then, after some algebra, we obtain

$$\begin{aligned} &b(\dot u_1(t) - \dot u_2(t), \dot u_1(t) - \dot u_2(t)) \\ &\quad \le a(u_1(t) - u_2(t), \dot u_2(t) - \dot u_1(t)) + j(Su_1(t), \dot u_2(t)) - j(Su_1(t), \dot u_1(t)) \\ &\qquad + j(Su_2(t), \dot u_1(t)) - j(Su_2(t), \dot u_2(t)). \end{aligned}$$

Next, using the assumptions (4.3), (4.4), (3.54), and (4.40), we find

$$\|\dot u_1(t) - \dot u_2(t)\|_X \le c\left(\|u_1(t) - u_2(t)\|_X + \int_0^t \|u_1(s) - u_2(s)\|_X\, ds \right). \tag{4.51}$$

Moreover, since $u_i(0) = u_0$, it follows that

$$u_i(t) = \int_0^t \dot u_i(s)\, ds + u_0 \quad \forall\, i = 1,2 \tag{4.52}$$

and, therefore,

$$\|u_1(t) - u_2(t)\|_X \le \int_0^t \|\dot u_1(s) - \dot u_2(s)\|_X\, ds. \tag{4.53}$$

We combine now inequalities (4.51) and (4.53), and use Gronwall's inequality (Lemma 1.44) to deduce that $\dot u_1 = \dot u_2$. Using (4.52), we obtain that $u_1(t) = u_2(t)$, which concludes the uniqueness part of Theroem 4.9(1).

(2) The proof is similar to the proof of (1). The only difference consists in the choice of the operator $\Lambda : C([0,T]; X \times Y) \to C([0,T]; X \times Y)$, which is

now given by

$$\Lambda\eta(t) = \left(\int_0^t w_\eta(s)\,ds + u_0, Sw_\eta(t)\right)$$
$$\forall\,\eta \in C([0,T];X \times Y),\ t \in [0,T]. \qquad (4.54)$$

Using (4.40), it follows that inequality (4.48) still holds and, therefore, Λ has a unique fixed point η^*. Denote by $u \in C^1([0,T];X)$ the function given by (4.49). It follows that u is a solution to problem (4.38)–(4.39) and, moreover, it is the unique solution of this Cauchy problem. □

Theorem 4.9 provides existence and uniqueness results in the study of the Cauchy problems (4.36)–(4.37) and (4.38)–(4.39), respectively, under the same assumptions on the data; also, note that the regularity $u \in C^1(0,T];X)$ of the solution is the same, in both the two problems above. Therefore, we conclude that the quasivariational inequalities (4.36)–(4.37) and (4.38)–(4.39) present a common feature. Moreover, we note that in comparison with Theorem 4.7, in Theorem 4.9 we do not need smallness assumptions in the study of problem (4.38)–(4.39). This feature arises from the fact that in the proof of Theorem 4.9, we use assumption (4.40), which involves an integral term and allows us to derive inequality (4.48); and, as it follows from Lemma 1.42, inequalities of this form lead to a fixed point property without any smallness assumption on the data. Examples of boundary value problems that lead to quasivariational inequalities of the form (4.36)–(4.37) or (4.38)–(4.39) will be presented in Section 10.4 of the book.

Chapter 5
Evolutionary Variational Inequalities

In this chapter, we continue the study of evolutionary variational inequalities started in Chapter 4. The difference between the problems studied there and those studied here lies in the fact that the variational inequalities presented in this chapter do not involve a viscosity term. Their study is more complicated, since it cannot be done based on the unique solvability of time-dependent elliptic variational inequalities in velocities. We start with the study of evolutionary variational inequalities involving a differentiable functional j, for which we prove an existence and uniqueness result. The proof is based on the study of a sequence of evolutionary variational inequalities with viscosity, compactness, and lower semicontinuity arguments. Then, we extend this result to a class of evolutionary variational inequalities for which the function j can be approached, in a sense that will be described below, by a family of differentiable functionals. We complete our results with a convergence result that shows that the solution of the evolutionary variational inequality with viscosity converges to the solution of the corresponding inviscid evolutionary variational inequality, as the viscosity converges to zero. Finally, we present an existence result for evolutionary quasivariational inequalities, obtained by using a time discretization method. The results presented in this chapter will be applied in the study of quasistatic antiplane frictional contact problems with elastic materials. As usual, everywhere in this chapter X is a real Hilbert space with the inner product $(\cdot, \cdot)_X$ and the norm $\|\cdot\|_X$, and $[0, T]$ denotes the time interval of interest, $T > 0$. Moreover, X is assumed to be separable everywhere in Section 5.4 of this chapter.

5.1 A First Existence and Uniqueness Result

Let a be a bilinear form on X, $j : X \to (-\infty, \infty]$, $f : [0, T] \to X$ and let u_0 be the initial data. We are interested in providing sufficient conditions that guarantee the existence of a unique solution of the following problem: find

$u : [0, T] \to X$ such that

$$a(u(t), v - \dot{u}(t)) + j(v) - j(\dot{u}(t)) \geq (f(t), v - \dot{u}(t))_X \qquad \forall\, v \in X, \ \text{a.e. } t \in (0, T), \tag{5.1}$$

$$u(0) = u_0. \tag{5.2}$$

Clearly, (5.1) represents an *evolutionary variational inequality* since it involves the time derivative of the unknown u and, as a consequence, the problem above involves an initial condition, (5.2). Note that as compared with problem (4.1)–(4.2) studied in the previous chapter, (5.1)–(5.2) does not involve a viscosity term, however, (5.1) can be obtained, formally, as a limit case of (4.1) when the viscosity term $b(\cdot, \cdot)$ vanishes. This property will be clearly described in Section 5.3 where we shall present a convergence result, Theorem 5.14; also, it will be used below in this section, in the proof of Theorem 5.1.

Evolutionary variational inequalities of the form (5.1)–(5.2) have been studied by many authors, by using different functional methods. Details can be found in the Bibliographical Notes at the end of this part of the book. Some of them used arguments of nonlinear equations with maximal monotone operators, and others employed a time discretization method. Here we prove an existence and uniqueness result by using a different method. Even if the results we present below are obtained under quite restrictive assumptions on the function j, they present some interest in their own right and are sufficiently general to provide appropriate mathematical tools in the study of the quasistatic frictional contact problems presented in Section 9.3 of this book.

In the study of the Cauchy problem (5.1)–(5.2), we assume that

$$\left.\begin{array}{l} a : X \times X \to \mathbb{R} \text{ is a bilinear symmetric form and} \\ \text{(a) there exists } M > 0 \text{ such that} \\ \qquad |a(u, v)| \leq M \, \|u\|_X \, \|v\|_X \quad \forall\, u, v \in X. \\ \text{(b) there exists } m > 0 \text{ such that} \\ \qquad a(v, v) \geq m \, \|v\|_X^2 \quad \forall\, v \in X. \end{array}\right\} \tag{5.3}$$

$$\left.\begin{array}{l} j : X \to \mathbb{R} \text{ is a convex Gâteaux differentiable functional and} \\ \text{(a) } j(v) \geq 0 \quad \forall\, v \in X. \\ \text{(b) } j(0_X) = 0. \\ \text{(c) there exists } c_j > 0 \text{ such that} \\ \qquad \|\nabla j(v)\|_X \leq c_j \left(\|v\|_X + 1\right) \quad \forall\, v \in X. \end{array}\right\} \tag{5.4}$$

$$f \in W^{1,2}(0, T; X). \tag{5.5}$$

$$u_0 \in X. \tag{5.6}$$

$$\left.\begin{array}{l} \text{There exists } \delta_0 \geq 0 \text{ such that} \\ a(u_0, v) + j(v) \geq (f(0), v)_X - \delta_0 \quad \forall\, v \in X. \end{array}\right\} \tag{5.7}$$

Note that assumption (5.4) combined with Corollary 2.33 provides sufficient conditions for the existence of the integrals below in this section, in which the functional j or its gadient ∇j appear.

Condition (5.7) represents a compatibility condition for the initial data of problem (5.1)–(5.2). Such kind of conditions are usual in the study of inviscid evolutionary variational inequalities. On occasions, assumption (5.7) is considered with $\delta_0 = 0$, as we shall do in Section 9.3 where we provide a mechanical interpretation of this assumption.

Our main result in this section is the following.

Theorem 5.1. *Let X be a Hilbert space and assume that* (5.3)–(5.7) *hold. Then there exists a unique solution u to problem* (5.1)–(5.2) *and it satisfies $u \in W^{1,2}(0,T;X)$. Moreover, there exists a positive constant c, which does not depend on j, such that*

$$\|u\|_{W^{1,2}(0,T;X)} \leq c\left(\sqrt{\delta_0} + \|u_0\|_X + \|f\|_{W^{1,2}(0,T;X)}\right). \tag{5.8}$$

The proof of Theorem 5.1 will be carried out in several steps and is based on the study of a sequence of evolutionary variational inequalities with viscosity, compactness, and lower semicontinuity arguments. We suppose in what follows that (5.3)–(5.7) hold, and, for every $n \in \mathbb{N}$, we consider the problem of finding $u_n : [0,T] \to X$ such that

$$\begin{aligned} a(u_n(t), v - \dot{u}_n(t)) + \frac{1}{n}\,(\dot{u}_n(t), v - \dot{u}_n(t))_X + j(v) - j(\dot{u}_n(t)) \\ \geq (f(t), v - \dot{u}_n(t))_X \quad \forall\, v \in X,\ t \in [0,T]. \end{aligned} \tag{5.9}$$

$$u_n(0) = u_0. \tag{5.10}$$

Below in this section, c denotes a positive generic constant that may depend on a, T and on the function φ, which will be introduced later, but does not depend on n and j, nor on time, and whose value may change from line to line.

In the first step, we provide the solvability of the problem (5.9)–(5.10), which represents an evolutionary variational inequality with viscosity.

Lemma 5.2. *For every $n \in \mathbb{N}$, there exists a unique solution u_n to the problem* (5.9)–(5.10). *Moreover, the solution satisfies $u_n \in W^{2,2}(0,T;X)$.*

Proof. Let $n \in \mathbb{N}$ and denote $b_n(u,v) = \frac{1}{n}\,(u,v)_X$ for all u, $v \in X$. It is clear that the form b_n satisfies (4.4) and, moreover, (5.4) and Corollary 1.37 imply that the function j is convex and lower semicontinuous. Lemma 5.2 represents now a consequence of Theorems 4.1 and 4.5. □

In the next step, we provide some estimates of the solutions u_n on the interval $[0,T]$.

Lemma 5.3. *There exists $c > 0$, which does not depend on j, such that the inequalities below hold, for all $n \in \mathbb{N}$:*

$$\|u_n\|_{C([0,T];X)} \leq c\,(\|u_0\|_X + \|f\|_{W^{1,2}(0,T;X)}), \tag{5.11}$$

$$\frac{1}{\sqrt{n}}\,\|\dot{u}_n\|_{L^2(0,T;X)} \leq c\,(\|u_0\|_X + \|f\|_{W^{1,2}(0,T;X)}), \tag{5.12}$$

$$\int_0^T j(\dot{u}_n(s))\,ds \leq c\,(\|u_0\|_X^2 + \|f\|^2_{W^{1,2}(0,T;X)}). \tag{5.13}$$

Proof. Let $n \in \mathbb{N}$. We use (5.4) to see that the evolutionary variational inequality (5.9) is equivalent to the variational equation

$$\begin{aligned} a(u_n(t), v) + \frac{1}{n}\,(\dot{u}_n(t), v)_X + (\nabla j(\dot{u}_n(t)), v)_X \\ = (f(t), v)_X \quad \forall\, v \in X,\ t \in [0,T] \end{aligned} \tag{5.14}$$

where ∇j denotes the gradient of the functional j, see Definition 1.35, page 14.

Let $s \in [0,T]$; we choose $v = \dot{u}_n(t)$ in (5.14) and integrate the resulting equality on $[0,s]$ to obtain

$$\begin{aligned} \int_0^s a(u_n(t), \dot{u}_n(t))\,dt + \frac{1}{n}\int_0^s (\dot{u}_n(t), \dot{u}_n(t))_X\,dt + \int_0^s (\nabla j(\dot{u}_n(t)), \dot{u}_n(t))_X\,dt \\ = \int_0^s (f(t), \dot{u}_n(t))_X\,dt. \end{aligned} \tag{5.15}$$

Next,

$$\begin{aligned} \int_0^s a(u_n(t), \dot{u}_n(t))\,dt &= \frac{1}{2}\int_0^s \frac{d}{dt}\,a(u_n(t), u_n(t))\,dt \\ &= \frac{1}{2}\,a(u_n(s), u_n(s)) - \frac{1}{2}\,a(u_0, u_0), \end{aligned} \tag{5.16}$$

$$\begin{aligned} &\int_0^s (f(t), \dot{u}_n(t))_X\,dt \\ &\quad = (f(s), u_n(s))_X - (f(0), u_n(0))_X - \int_0^s (\dot{f}(t), u_n(t))_X\,dt \end{aligned} \tag{5.17}$$

and

$$\int_0^s (\nabla j(\dot{u}_n(t)), \dot{u}_n(t))_X\,dt \geq \int_0^s j(\dot{u}_n(t))\,dt, \tag{5.18}$$

where the inequality is obtained from the subgradient inequality

$$j(v) - j(\dot{u}_n(t)) \geq (\nabla j(\dot{u}_n(t)), v - \dot{u}_n(t)) \quad \forall\, v \in X$$

by choosing $v = 0_X$, using the assumption that $j(0_X) = 0$ and integrating the result on $[0,s]$.

We now combine (5.15)–(5.18) and use (5.3) and the initial condition $u_n(0) = u_0$; as a result, after some algebra we find

$$\frac{m}{2}\|u_n(s)\|_X^2 + \frac{1}{n}\int_0^s \|\dot{u}_n(t)\|_X^2\,dt + \int_0^s j(\dot{u}_n(t))\,dt$$
$$\leq c(\|u_0\|_X^2 + \|f(0)\|_X^2) + \|f(s)\|_X\|u_n(s)\|_X + \int_0^s \|\dot{f}(t)\|_X\|u_n(t)\|_X\,dt. \tag{5.19}$$

We now use the inequality

$$ab \leq \frac{\alpha\, a^2}{2} + \frac{b^2}{2\,\alpha} \quad \text{for } a,\, b,\, \alpha > 0$$

with a convenient choice of α to see that (5.19) leads to

$$\|u_n(s)\|_X^2 + \frac{1}{n}\int_0^s \|\dot{u}_n(t)\|_X^2\,dt + \int_0^s j(\dot{u}_n(t))\,dt$$
$$\leq c\,(\|u_0\|_X^2 + \|f\|_{W^{1,2}(0,T;X)}^2) + \int_0^s \|u_n(t)\|_X^2\,dt. \tag{5.20}$$

Since $j \geq 0$, (5.4)(a), it follows from (5.20) that

$$\|u_n(s)\|_X^2 \leq c\,(\|u_0\|_X^2 + \|f\|_{W^{1,2}(0,T;X)}^2) + \int_0^s \|u_n(t)\|_X^2\,dt,$$

and, therefore, by using Gronwall's inequality yields

$$\|u_n(s)\|_X^2 \leq c\,(\|u_0\|_X^2 + \|f\|_{W^{1,2}(0,T;X)}^2), \tag{5.21}$$

which implies (5.11).

We use again (5.20) and (5.4)(a) to see that

$$\frac{1}{n}\int_0^s \|\dot{u}_n(t)\|_X^2\,dt \leq c\,(\|u_0\|_X^2 + \|f\|_{W^{1,2}(0,T;X)}^2) + \int_0^s \|u_n(t)\|_X^2\,dt$$

and, by (5.11), we find

$$\frac{1}{n}\,\|\dot{u}_n\|_{L^2(0,T;X)}^2 \leq c\,(\|u_0\|_X^2 + \|f\|_{W^{1,2}(0,T;X)}^2)$$

which implies (5.12).

Finally, we combine (5.20) and (5.11) to obtain (5.13), which concludes the proof of the lemma. □

We now provide an estimate on the solutions u_n on the interval $[0, T_0]$, where $0 < T_0 < T$.

Lemma 5.4. *For all $0 < T_0 < T$, there exists $c(T_0) > 0$, which does not depend on j, such that*

$$\|\ddot{u}_n\|_{L^2(0,T_0;X)} \leq c(T_0)(\sqrt{\delta_0} + \|u_0\|_X + \|f\|_{W^{1,2}(0,T;X)}), \tag{5.22}$$

for all $n \in \mathbb{N}$.

Proof. Let $n \in \mathbb{N}$ and consider a Lipschitz continuous function $\varphi : [0,T] \to \mathbb{R}$ such that $\varphi(t) > 0$ for all $t \in [0,T)$ and $\varphi(T) = 0$. We take $v = \varphi(t)\ddot{u}_n(t)$ in (5.14) and integrate the result on $[0,T]$ to obtain

$$\begin{aligned}\int_0^T \varphi(t)\, a(u_n(t), \ddot{u}_n(t))\, dt + \frac{1}{n}\int_0^T \varphi(t)\, (\dot{u}_n(t), \ddot{u}_n(t))_X\, dt \\ + \int_0^T \varphi(t)\, (\nabla j(\dot{u}_n(t)), \ddot{u}_n(t))_X\, dt = \int_0^T \varphi(t)\, (f(t), \ddot{u}_n(t))_X\, dt.\end{aligned} \tag{5.23}$$

We now perform integrations by parts and use the assumption $\varphi(T) = 0$ to see that

$$\begin{aligned}&\int_0^T \varphi(t)\, a(u_n(t), \ddot{u}_n(t))\, dt \\ &\quad = -\varphi(0)\, a(u_n(0), \dot{u}_n(0)) - \int_0^T \varphi(t)\, a(\dot{u}_n(t), \dot{u}_n(t))\, dt \\ &\qquad - \int_0^T \dot{\varphi}(t)\, a(u_n(t), \dot{u}_n(t))\, dt,\end{aligned} \tag{5.24}$$

$$\begin{aligned}&\frac{1}{n}\int_0^T \varphi(t)\, (\dot{u}_n(t), \ddot{u}_n(t))_X\, dt \\ &\quad = -\frac{1}{2n}\left[\varphi(0)\, (\dot{u}_n(0), \dot{u}_n(0))_X + \int_0^T \dot{\varphi}(t)\, (\dot{u}_n(t), \dot{u}_n(t))_X\, dt\right],\end{aligned} \tag{5.25}$$

$$\begin{aligned}&\int_0^T \varphi(t)\, (\nabla j(\dot{u}_n(t)), \ddot{u}_n(t))_X\, dt \\ &\quad = \int_0^T \varphi(t)\, \frac{d}{dt} j(\dot{u}_n(t))\, dt = -\varphi(0)\, j(\dot{u}_n(0)) - \int_0^T \dot{\varphi}(t)\, j(\dot{u}_n(t))\, dt,\end{aligned} \tag{5.26}$$

$$\begin{aligned}&\int_0^T \varphi(t)\, (f(t), \ddot{u}_n(t))_X\, dt \\ &\quad = -\varphi(0)\, (f(0), \dot{u}_n(0))_X - \int_0^T \varphi(t)\, (\dot{f}(t), \dot{u}_n(t))_X\, dt \\ &\qquad - \int_0^T \dot{\varphi}(t)\, (f(t), \dot{u}_n(t))_X\, dt.\end{aligned} \tag{5.27}$$

We substitute equalities (5.24)–(5.27) in (5.23) and use the initial condition (5.10), assumption (5.7) with $v = \dot{u}_n(0)$, and inequality $\varphi(0) > 0$ to obtain

$$\begin{aligned}
&\int_0^T \varphi(t)\, a(\dot{u}_n(t), \dot{u}_n(t))\, dt \\
&\quad \le \delta_0 \varphi(0) - \frac{1}{2n} \int_0^T \dot{\varphi}(t)\, \|\dot{u}_n(t)\|_X^2\, dt - \int_0^T \dot{\varphi}(t)\, j(\dot{u}_n(t))\, dt \\
&\qquad - \int_0^T \dot{\varphi}(t)\, a(u_n(t), \dot{u}_n(t))\, dt + \int_0^T \varphi(t)\, (\dot{f}(t), \dot{u}_n(t))_X\, dt \\
&\qquad + \int_0^T \dot{\varphi}(t)\, (f(t), \dot{u}_n(t))_X\, dt. \qquad (5.28)
\end{aligned}$$

On the other hand, we use (5.12) to see that

$$-\frac{1}{2n} \int_0^T \dot{\varphi}(t) \|\dot{u}_n(t)\|_X^2\, dt \le c\,(\|u_0\|_X^2 + \|f\|_{W^{1,2}(0,T;X)}^2) \qquad (5.29)$$

and we recall that here and below, c is a positive constant that depends on φ but is independent of j and n and whose value may change from place to place.

Also, inequality (5.13) implies that

$$-\int_0^T \dot{\varphi}(t)\, j(\dot{u}_n(t))\, dt \le c\,(\|u_0\|_X^2 + \|f\|_{W^{1,2}(0,T;X)}^2). \qquad (5.30)$$

Assume now that $\frac{|\dot{\varphi}|}{\sqrt{\varphi}} \in L^\infty(0,T)$ and note that this additional assumption is compatible with the previous assumption on φ since, for example, the function $\varphi(t) = (T-t)^2$ satisfies all these requirements. With this additional assumption on the function φ, we use (5.3) to see that

$$\begin{aligned}
-\int_0^T \dot{\varphi}(t)\, a(u_n(t), \dot{u}_n(t))\, dt &\le c \int_0^T \sqrt{\varphi(t)}\, \|\dot{u}_n(t)\|_X \|u_n(t)\|_X \frac{|\dot{\varphi}(t)|}{\sqrt{\varphi(t)}}\, dt \\
&\le c \int_0^T \sqrt{\varphi(t)}\, \|\dot{u}_n(t)\|_X \|u_n(t)\|_X\, dt
\end{aligned}$$

and, using the inequality

$$ab \le \frac{\alpha\, a^2}{2} + \frac{b^2}{2\,\alpha} \quad \text{for } a,\ b,\ \alpha > 0,$$

we find that

$$-\int_0^T \dot{\varphi}(t) a(u_n(t), \dot{u}_n(t)) dt \le c\Big(\alpha \int_0^T \varphi(t) \|\dot{u}_n(t)\|_X^2 dt + \frac{1}{\alpha} \int_0^T \|u_n(t)\|_X^2 dt\Big), \qquad (5.31)$$

for all $\alpha > 0$. Using similar arguments, it follows that

$$\int_0^T \varphi(t)\,(\dot{f}(t), \dot{u}_n(t))_X\,dt \le c\left(\alpha \int_0^T \varphi(t)\,\|\dot{u}_n(t)\|_X^2\,dt + \frac{1}{\alpha}\int_0^T \|\dot{f}(t)\|_X^2\,dt\right) \tag{5.32}$$

and

$$\int_0^T \dot{\varphi}(t)\,(f(t), \dot{u}_n(t))_X\,dt \le c\left(\alpha \int_0^T \varphi(t)\,\|\dot{u}_n(t)\|_X^2\,dt + \frac{1}{\alpha}\int_0^T \|f(t)\|_X^2\,dt\right), \tag{5.33}$$

for all $\alpha > 0$. We combine now (5.28)–(5.33), use (5.3), and choose a sufficiently small α to obtain that

$$\int_0^T \varphi(t)\,\|\dot{u}_n(t)\|_X^2\,dt \le c\left(\delta_0 + \|u_0\|_X^2 + \|f\|_{W^{1,2}(0,T;X)}^2 + \int_0^T \|u_n(t)\|_X^2\,dt\right).$$

We use now the previous inequality and (5.11) to see that

$$\int_0^T \varphi(t)\,\|\dot{u}_n(t)\|_X^2 dt \le c\,(\delta_0 + \|u_0\|_X^2 + \|f\|_{W^{1,2}(0,T;X)}^2). \tag{5.34}$$

Let T_0 be such that $0 < T_0 < T$ and choose φ such that the lower bound of φ on $[0, T_0]$, denoted $l(T_0)$, satisfies

$$l(T_0) = \inf_{t\in[0,T_0]} \varphi(t) > 0. \tag{5.35}$$

For instance, take $\varphi(t) = (T-t)^2$ to obtain $l(T_0) = (T-T_0)^2 > 0$. Note also that $l(T_0)$ depends on T_0 but is independent on j and n. It follows from (5.34) and (5.35) that

$$\begin{aligned} l(T_0)\int_0^{T_0} \|\dot{u}_n(t)\|_X^2\,dt &\le \int_0^{T_0} \varphi(t)\,\|\dot{u}_n(t)\|_X^2\,dt \\ &\le \int_0^{T} \varphi(t)\,\|\dot{u}_n(t)\|_X^2\,dt \\ &\le c\,(\delta_0 + \|u_0\|_X^2 + \|f\|_{W^{1,2}(0,T;X)}^2) \\ &\le c\,(\sqrt{\delta_0} + \|u_0\|_X + \|f\|_{W^{1,2}(0,T;X)})^2. \end{aligned}$$

This last inequality leads to (5.22) with $c(T_0) = \sqrt{\frac{c}{l(T_0)}}$, which depends on T_0 but is independent on j and n. □

Inequality (5.11) shows that $\{u_n\}$ is a bounded sequence in the space $L^\infty(0,T;X)$, and inequality (5.22) shows that, for all $0 < T_0 < T$, $\{\dot{u}_n\}$

is a bounded sequence in the space $L^2(0,T_0;X)$. Therefore, there exists an element $u \in L^\infty(0,T;X)$ such that for all $0 < T_0 < T$, the derivative of u belongs to the space $L^2(0,T_0;X)$ and there exists a subsequence of the sequence $\{u_n\}$, again denoted $\{u_n\}$, such that

$$u_n \rightharpoonup^* u \quad \text{in } L^\infty(0,T;X), \tag{5.36}$$

$$\dot{u}_n \rightharpoonup \dot{u} \quad \text{in } L^2(0,T_0;X). \tag{5.37}$$

The convergences (5.36) and (5.37) imply that

$$u_n(t) \rightharpoonup u(t) \quad \text{in } X, \quad \text{for all } t \in [0,T_0]. \tag{5.38}$$

In the next lemma, we reinforce (5.38) by showing that, in fact, the sequence $\{u_n(t)\}$ convergences strongly in X and uniformly with respect to $t \in [0,T_0]$. Although this last result is not really necessary to the proof of Theorem 5.1, we present it since it has interest in and of itself, and it is related to the convergence result presented in Section 5.3.

Lemma 5.5. *Let $0 < T_0 < T$ and consider a subsequence of the sequence $\{u_n\}$, again denoted $\{u_n\}$, which satisfies* (5.36) *and* (5.37). *Then*

$$\|u_n - u\|_{C([0,T_0];X)} \to 0 \quad \text{as } n \to \infty. \tag{5.39}$$

Proof. Let $t \in [0,T]$ and let $p,\ n \in \mathbb{N}$ such that $p > n$. It follows from (5.9) that

$$a(u_n(t), \dot{u}_p(t) - \dot{u}_n(t)) + \frac{1}{n}\,(\dot{u}_n(t), \dot{u}_p(t) - \dot{u}_n(t))_X + j(\dot{u}_p(t)) - j(\dot{u}_n(t))$$
$$\geq (f(t), \dot{u}_p(t) - \dot{u}_n(t))_X$$

and

$$a(u_p(t), \dot{u}_n(t) - \dot{u}_p(t)) + \frac{1}{p}\,(\dot{u}_p(t), \dot{u}_n(t) - \dot{u}_p(t))_X + j(\dot{u}_n(t)) - j(\dot{u}_p(t))$$
$$\geq (f(t), \dot{u}_n(t) - \dot{u}_p(t))_X.$$

We add the two inequalities to obtain

$$\begin{aligned}
&a(u_n(t) - u_p(t), \dot{u}_n(t) - \dot{u}_p(t))\\
&\quad\leq \left(\frac{1}{p}\,\dot{u}_p(t) - \frac{1}{n}\,\dot{u}_n(t), \dot{u}_n(t) - \dot{u}_p(t)\right)_X\\
&\quad\leq \left(\frac{1}{p}\,\|\dot{u}_p(t)\|_X + \frac{1}{n}\,\|\dot{u}_n(t)\|_X\right)(\|\dot{u}_n(t)\|_X + \|\dot{u}_p(t)\|_X)\\
&\quad\leq \frac{2}{n}\,(\|\dot{u}_n(t)\|_X^2 + \|\dot{u}_p(t)\|_X^2).
\end{aligned}$$

Let $s \in [0, T_0]$. We integrate the previous inequality on $[0, s]$ with the initial conditions $u_n(0) = u_p(0) = u_0$ and use Proposition 2.30 to find that

$$a(u_n(s) - u_p(s), u_n(s) - u_p(s)) \le \frac{4}{n} \int_0^s (\|\dot{u}_p(t)\|_X^2 + \|\dot{u}_n(t)\|_X^2)\, dt. \quad (5.40)$$

We use now in (5.40) the X-ellipticity of the form a, (5.3)(b), and the estimate (5.22) derived in Lemma 5.4. As a result, we deduce that

$$\|u_n(s) - u_p(s)\|_X^2 \le \frac{k}{n}, \quad (5.41)$$

where k is a positive constant that depends on a, u_0, f, δ_0, φ, and T_0. Since s is an arbitrary point in $[0, T_0]$, it follows from (5.41) that

$$\|u_n - u_p\|_{C([0,T_0];X)} \le \sqrt{\frac{k}{n}},$$

i.e., $\{u_n\}$ is a Cauchy sequence in the Banach space $C([0, T_0]; X)$. Therefore, there exists a function $\tilde{u} \in C([0, T_0]; X)$ such that

$$\|u_n - \tilde{u}\|_{C([0,T_0];X)} \to 0 \quad \text{as } n \to \infty \quad (5.42)$$

which implies that

$$u_n \rightharpoonup^* \tilde{u} \ \text{ in } L^\infty(0, T_0; X). \quad (5.43)$$

It follows from (5.36) and (5.43) that $u = \tilde{u}$ on $[0, T_0]$ and, combining this last equality with (5.42), we deduce (5.39), which concludes the proof of the lemma. □

In the next two steps, we prove the unique local solvability of the evolutionary variational inequality (5.1)–(5.2) together with two additional estimates.

Lemma 5.6. *There exists a unique function* $u \in L^2(0, T; X)$ *such that* $u(0) = u_0$ *and, for all* $T_0 \in (0, T)$,

$$u \in W^{1,2}(0, T_0; X), \quad (5.44)$$

$$a(u(t), v - \dot{u}(t)) + j(v) - j(\dot{u}(t)) \ge (f(t), v - \dot{u}(t))_X \quad \forall v \in X, \ a.e.\ t \in (0, T_0). \quad (5.45)$$

Proof. Existence. Let $0 < T_0 < T$. We proved on page 83 that there exists an element u such that $u \in L^2(0, T; X)$, $\dot{u} \in L^2(0, T_0; X)$ and, moreover, there exists a subsequence of the sequence $\{u_n\}$, again denoted $\{u_n\}$, such that (5.36), (5.37), and (5.39) hold. Therefore we conclude that u satisfies (5.44).

Consider now an element $v \in L^2(0, T_0; X)$; we use $v(t)$ as the test function in (5.9) and integrate the result on $[0, T_0]$ to obtain, for all $n \in \mathbb{N}$,

$$\int_0^{T_0} a(u_n(t), v(t) - \dot{u}_n(t))\, dt + \frac{1}{n} \int_0^{T_0} (\dot{u}_n(t), v(t) - \dot{u}_n(t))_X\, dt + \int_0^{T_0} j(v(t))dt$$
$$\geq \int_0^{T_0} j(\dot{u}_n(t))dt + \int_0^{T_0} (f(t), v(t) - \dot{u}_n(t))_X\, dt. \tag{5.46}$$

It follows from (5.37), (5.39), and (5.3) that

$$\lim_{n\to\infty} \int_0^{T_0} a(u_n(t), v(t) - \dot{u}_n(t))\, dt = \int_0^{T_0} a(u(t), v(t) - \dot{u}(t))\, dt \tag{5.47}$$

and using (5.22) yields

$$\lim_{n\to\infty} \frac{1}{n} \int_0^{T_0} (\dot{u}_n(t), v(t) - \dot{u}_n(t))_X\, dt = 0. \tag{5.48}$$

Also, assumption (5.4) on the function j, Corollary 2.33, and (5.37) imply that

$$\liminf_{n\to\infty} \int_0^{T_0} j(\dot{u}_n(t))\, dt \geq \int_0^{T_0} j(\dot{u}(t))\, dt \tag{5.49}$$

and, again, (5.37) leads to

$$\lim_{n\to\infty} \int_0^{T_0} (f(t), v(t) - \dot{u}_n(t))_X dt = \int_0^{T_0} (f(t), v(t) - \dot{u}(t))_X\, dt. \tag{5.50}$$

We pass to the lower limit in (5.46) and use (5.47)–(5.50) to obtain

$$\int_0^{T_0} a(u(t), v(t) - \dot{u}(t))\, dt + \int_0^{T_0} j(v(t))dt - \int_0^{T_0} j(\dot{u}(t))\, dt$$
$$\geq \int_0^{T_0} (f(t), v(t) - \dot{u}(t))_X\, dt.$$

Since the previous inequality is valid for all $v \in L^2(0, T_0; X)$, a classical application of the Lebesgue point for L^1 functions (Theorem 2.20 on page 34, see also [57, p. 165]) shows that u satisfies (5.45).

Finally, (5.10) and (5.39) imply that $u(0) = u_0$, which concludes the existence part of the theorem.

Uniqueness. Consider two functions $u_1, u_2 \in L^2(0, T; X)$ such that $u_1(0) = u_2(0) = u_0$, which satisfy (5.44), (5.45), for all $0 < T_0 < T$. Let $s \in (0, T)$ and let T_0 be such that $s < T_0 < T$. We have

$$a(u_1(t), v - \dot{u}_1(t)) + j(v) - j(\dot{u}_1(t)) \geq (f(t), v - \dot{u}_1(t))_X,$$
$$a(u_2(t), v - \dot{u}_2(t)) + j(v) - j(\dot{u}_2(t)) \geq (f(t), v - \dot{u}_2(t))_X,$$

for all $v \in X$, a.e. $t \in (0, T_0)$. We take $v = \dot{u}_2(t)$ in the first inequality, $v = \dot{u}_1(t)$ in the second one, add the resulting inequalities, and use Proposition 2.30 to obtain

$$\frac{1}{2}\frac{d}{dt}\,a(u_1(t) - u_2(t), u_1(t) - u_2(t)) \le 0 \quad \text{a.e. } t \in (0, T_0).$$

We integrate the previous inequality on $[0, s]$ with the initial conditions $u_1(0) = u_2(0) = u_0$ and use assumption (5.3)(b) on a to see that

$$m\,\|u_1(s) - u_2(s)\|_X^2 \le 0.$$

This inequality shows that $u_1(s) = u_2(s)$ for all $s \in (0, T)$, i.e., $u_1 = u_2$. □

Lemma 5.7. *Let u be the function defined in Lemma 5.6. Then:*
(1) There exists $c > 0$, which does not depend on j, such that

$$\|u\|_{L^2(0,T;X)} \le c\,(\|u_0\|_X + \|f\|_{W^{1,2}(0,T;X)}). \tag{5.51}$$

(2) For all $0 < T_0 < T$, there exists $c(T_0) > 0$, which does not depend on j, such that

$$\|\dot{u}\|_{L^2(0,T_0;X)} \le c(T_0)(\sqrt{\delta_0} + \|u_0\|_X + \|f\|_{W^{1,2}(0,T;X)}). \tag{5.52}$$

Proof. (1) We use (5.36) to see that $u_n \rightharpoonup u$ weakly in $L^2(0, T; X)$ and, since the arguments presented on page 13 show that $v \to \|v\|_{L^2(0,T;X)}$ is a l.s.c. functional on the space $L^2(0, T; X)$, we find that

$$\|u\|_{L^2(0,T;X)} \le \liminf_{n\to\infty} \|u_n\|_{L^2(0,T;X)}. \tag{5.53}$$

Also, since $\|u_n\|_{L^2(0,T;X)} \le \sqrt{T}\,\|u_n\|_{C([0,T];X)}$ for all $n \in \mathbb{N}$, it follows from (5.11) that

$$\|u_n\|_{L^2(0,T;X)} \le c\,(\|u_0\|_X + \|f\|_{W^{1,2}(0,T;X)}) \quad \forall n \in \mathbb{N} \tag{5.54}$$

where, again, c is a positive constant that does not depend on n and j. Estimate (5.51) is now a consequence of (5.53) and (5.54).

(2) Let $0 < T_0 < T$. We use (5.37), (5.22), and the lower semicotinuity of the norm on the space $L^2(0, T_0; X)$ to see that

$$\begin{aligned}\|\dot{u}\|_{L^2(0,T_0;X)} &\le \liminf_{n\to\infty} \|\dot{u}_n\|_{L^2(0,T_0;X)} \\ &\le c(T_0)(\sqrt{\delta_0} + \|u_0\|_X + \|f\|_{W^{1,2}(0,T;X)}),\end{aligned}$$

which concludes the proof. □

We have now all the ingredients needed to prove Theorem 5.1.

Proof. Existence. Let $T_1 > T$; we extend the function $f : [0,T] \to X$ to a function $f_1 : [0,T_1] \to X$ given by

$$f_1(t) = \begin{cases} f(t) & \text{if } 0 \le t \le T, \\ f(T) & \text{if } T < t \le T_1. \end{cases} \tag{5.55}$$

We note that (5.5) implies the regularity $f_1 \in W^{1,2}(0,T_1;X)$, and (5.7) still holds if we replace $f(0)$ by $f_1(0)$. Therefore, we are allowed to apply Lemmas 5.6 and 5.7 on the time interval $(0,T_1)$, replacing f with f_1; as a result, we obtain that there exists a unique function $u_1 \in L^2(0,T_1;X)$ such that

$$u_1(0) = u_0 \tag{5.56}$$

and, for all $T_0 \in (0,T_1)$,

$$u_1 \in W^{1,2}(0,T_0;X), \tag{5.57}$$

$$\begin{aligned} a(u_1(t), v - \dot{u}_1(t)) + j(v) - j(\dot{u}_1(t)) \\ \ge (f_1(t), v - \dot{u}_1(t))_X \quad \forall\, v \in X, \text{ a.e. } t \in (0,T_0), \end{aligned} \tag{5.58}$$

$$\|\dot{u}_1\|_{L^2(0,T_0;X)} \le c(T_0)(\sqrt{\delta_0} + \|u_0\|_X + \|f_1\|_{W^{1,2}(0,T_1;X)}), \tag{5.59}$$

where $c(T_0) > 0$ does not depend on j.

Let u^* be the restriction of u_1 to $[0,T]$, i.e.,

$$u^*(t) = u_1(t) \quad \forall\, t \in [0,T]. \tag{5.60}$$

We take $T_0 = T < T_1$ and use (5.56)–(5.58) to see that u^* is a solution of problem (5.1)–(5.2) and satisfies $u^* \in W^{1,2}(0,T;X)$.

Uniqueness. The uniqueness of the solution follows directly from (5.1)–(5.2), by using the same arguments as those used in the proof of the uniqueness part of Lemma 5.6.

Estimate. We use (5.59) with $T_0 = T < T_1$ and (5.60) to obtain that

$$\|\dot{u}^*\|_{L^2(0,T;X)} \le c(T)\big(\sqrt{\delta_0} + \|u_0\|_X + \|f_1\|_{W^{1,2}(0,T_1;X)}\big)$$

and, since (5.55) implies that

$$\|f_1\|_{W^{1,2}(0,T_1;X)} \le (1 + c\sqrt{T_1 - T}\ )\ \|f\|_{W^{1,2}(0,T;X)},$$

we find that

$$\|\dot{u}^*\|_{L^2(0,T;X)} \le c(T)\,(\sqrt{\delta_0} + \|u_0\|_X + (1 + c\sqrt{T_1 - T}\ )\ \|f\|_{W^{1,2}(0,T;X)}). \tag{5.61}$$

Also, it follows from (5.56)–(5.58) and (5.60) that $u^*(0) = u_0$ and, for all $T_0 \in (0,T)$, u^* satisfies (5.44) and (5.45). We use now the uniqueness part in Lemma 5.6 to see that $u^* = u$ where, recall, u is the function defined in this lemma. Combining this last equality with (5.51), we obtain that

$$\|u^*\|_{L^2(0,T;X)} \leq c\,(\|u_0\|_X + \|f\|_{W^{1,2}(0,T;X)}). \tag{5.62}$$

We now use inequalities (5.62) and (5.61) and take $T_1 = T+1$ to deduce that u^* satisfies an inequality of the form (5.8) in which c does not depend on j. □

5.2 Regularization

Our aim in this section is to solve the Cauchy problem (5.1)–(5.2) in the case when j is not a Gâteaux differentiable function, but it can be approached, in a sense that will be described below, by a family of Gâteaux differentiable functionals. To this end, we assume that (5.3), (5.5), (5.6) hold and, moreover,

$$\left.\begin{array}{l} j : X \to \mathbb{R} \text{ is a convex l.s.c. function and there exists} \\ c > 0 \text{ such that } j(v) \leq c\left(\|v\|_X^2 + 1\right) \ \forall\, v \in X. \end{array}\right\} \tag{5.63}$$

We also assume that there exists a family of functionals $(j_\rho)_{\rho>0}$ such that:

$$\text{For each } \rho > 0 \ \text{ the functional } j_\rho \text{ satisfies (5.4).} \tag{5.64}$$

$$\lim_{\rho\to 0} \int_0^T j_\rho(v(t))\,dt = \int_0^T j(v(t))\,dt \quad \forall\, v \in L^2(0,T;X). \tag{5.65}$$

$$\left.\begin{array}{l} \text{For each sequence } \{v_\rho\} \subset L^2(0,T;X) \text{ such that} \\ v_\rho \rightharpoonup v \quad \text{in} \quad L^2(0,T;X) \quad \text{as} \quad \rho \to 0, \\ \text{the inequality below holds :} \\ \displaystyle\liminf_{\rho\to 0} \int_0^T j_\rho(v_\rho(t))\,dt \geq \int_0^T j(v(t))\,dt. \end{array}\right\} \tag{5.66}$$

Note that the existence of the integrals in (5.65) and (5.65) is a consequence of conditions (5.63) and (5.64), as it results from Corollaries 2.32 and 2.33.

Finally, we assume the compatibility condition

$$\left.\begin{array}{l} \text{For each } \rho > 0 \text{ there exists } \delta_{0\rho} \geq 0 \text{ such that} \\ a(u_0, v) + j_\rho(v) \geq (f(0), v)_X - \delta_{0\rho} \quad \forall\, v \in X \end{array}\right\} \tag{5.67}$$

and we complete it by the following boundedness condition:

$$\left.\begin{array}{l}\text{There exist } \rho^* > 0 \text{ and } M^* > 0 \text{ such that}\\ \delta_{0\rho} < M^* \text{ for all } \rho \in (0, \rho^*).\end{array}\right\} \tag{5.68}$$

Note that this last condition shows that $\{\delta_{0\rho}\}$ is a bounded sequence as $\rho \to 0$.

For each $\rho > 0$, we consider the problem of finding a function $u_\rho : [0, T] \to X$ such that

$$a(u_\rho(t), v - \dot{u}_\rho(t)) + j_\rho(v) - j_\rho(\dot{u}_\rho(t)) \geq (f(t), v - \dot{u}_\rho(t))_X \quad \forall\, v \in X, \text{ a.e. } t \in (0, T), \tag{5.69}$$

$$u_\rho(0) = u_0. \tag{5.70}$$

Our first result in this section is the following.

Theorem 5.8. *Let X be a Hilbert space and assume that* (5.3), (5.5), (5.6), *and* (5.63)–(5.68) *hold. Then:*

(1) For each $\rho > 0$, there exists a unique solution to problem (5.2)–(5.70) *and it satisfies $u_\rho \in W^{1,2}(0, T; X)$.*

(2) There exists a unique solution to problem (5.1)–(5.2) *and it satisfies $u \in W^{1,2}(0, T; X)$.*

(3) The solution u_ρ of problem (5.2)–(5.70) *converges weakly to the solution u of problem* (5.1)–(5.2), *i.e.,*

$$u_\rho \rightharpoonup u \quad \text{in } W^{1,2}(0, T; X) \quad \text{as } \rho \to 0. \tag{5.71}$$

Proof. (1) Let $\rho > 0$. The existence of a unique function $u_\rho \in W^{1,2}(0, T; X)$ that satisfies (5.2)–(5.70) is a consequence of Theorem 5.1. Note that by Theorem 5.1, we also deduce that there exists a positive constant c, which does not depend on ρ, such that

$$\|u_\rho\|_{W^{1,2}(0,T;X)} \leq c\big(\sqrt{\delta_{0\rho}} + \|u_0\|_X + \|f\|_{W^{1,2}(0,T;X)}\big). \tag{5.72}$$

(2) We now investigate the behavior of the sequence $\{u_\rho\}$ as $\rho \to 0$ and, therefore, we assume in what follows that $0 < \rho < \rho^*$, where ρ^* is defined in (5.68). Estimate (5.72) combined with assumption (5.68) show that the sequence $\{u_\rho\}$ is bounded in $W^{1,2}(0, T; X)$; therefore, by using Theorem 1.21, we can find an element $u \in W^{1,2}(0, T; X)$ and a subsequence of $\{u_\rho\}$, again denoted $\{u_\rho\}$, for which (5.71) holds. The convergence (5.71) yields

$$u_\rho \rightharpoonup u \quad \text{in } L^2(0, T; X) \quad \text{as } \rho \to 0, \tag{5.73}$$

$$\dot{u}_\rho \rightharpoonup \dot{u} \quad \text{in } L^2(0, T; X) \quad \text{as } \rho \to 0. \tag{5.74}$$

Let $v \in L^2(0,T;X)$; we integrate (5.2) on $[0,T]$ with the initial condition (5.70) to obtain

$$\begin{aligned}\int_0^T a(u_\rho(t),v(t))\,dt + \int_0^T j_\rho(v(t))\,dt \\ \geq \frac{1}{2}\,a(u_\rho(T),u_\rho(T)) - \frac{1}{2}\,a(u_0,u_0) + \int_0^T j_\rho(\dot{u}_\rho(t))\,dt \\ + \int_0^T (f(t), v(t)-\dot{u}_\rho(t))_X\,dt. \end{aligned} \tag{5.75}$$

Then, we use (5.73), (5.74), the properties of the bilinear form a, (5.3), and assumptions (5.65), (5.65) to see that

$$\lim_{\rho\to 0} \int_0^T a(u_\rho(t),v(t))\,dt = \int_0^T a(u(t),v(t))\,dt, \tag{5.76}$$

$$\liminf_{\rho\to 0} \frac{1}{2}\,a(u_\rho(T),u_\rho(T)) \geq \frac{1}{2}\,a(u(T),u(T)), \tag{5.77}$$

$$\lim_{\rho\to 0} \int_0^T j_\rho(v(t))\,dt = \int_0^T j(v(t))\,dt, \tag{5.78}$$

$$\liminf_{\rho\to 0} \int_0^T j_\rho(\dot{u}_\rho(t))\,dt \geq \int_0^T j(\dot{u}(t))\,dt, \tag{5.79}$$

$$\lim_{\rho\to 0} \int_0^T (f(t),v(t)-\dot{u}_\rho(t))_X\,dt = \int_0^T (f(t),v(t)-\dot{u}(t))_X\,dt. \tag{5.80}$$

Next, we pass to the lower limit in (5.75) as $\rho \to 0$ and use (5.76)–(5.80) to obtain

$$\begin{aligned}\int_0^T a(u(t),v(t))\,dt + \int_0^T j(v(t))\,dt \\ \geq \frac{1}{2}\,a(u(T),u(T)) - \frac{1}{2}\,a(u_0,u_0) + \int_0^T j(\dot{u}(t))\,dt \\ + \int_0^T (f(t), v(t)-\dot{u}(t))_X\,dt. \end{aligned} \tag{5.81}$$

It follows from the convergences (5.73) and (5.74) that $u_\rho(t) \rightharpoonup u(t)$ in X for all $t \in [0,T]$ and, using (5.70), we deduce that u satisfies (5.2). This last equality and the properties of a combined with Proposition 2.30 on page 37 yield

$$\int_0^T a(u(t),\dot{u}(t))\,dt = \frac{1}{2}\,a(u(T),u(T)) - \frac{1}{2}\,a(u_0,u_0). \tag{5.82}$$

We now use (5.81) and (5.82) to see that u satisfies the inequality

$$\int_0^T a(u(t),v(t)-\dot{u}(t))\,dt + \int_0^T (j(v(t))-j(\dot{u}(t)))\,dt \geq \int_0^T (f(t),v(t)-\dot{u}(t))_X dt.$$

Then, performing a localization argument by applying Theorem 2.20 on page 34, we obtain that u satisfies (5.1).

We conclude that the element $u \in W^{1,2}(0,T;X)$ defined above is a solution to problem (5.1)–(5.2), which proves the existence of a solution to this Cauchy problem. The uniqueness of the solution follows by using the arguments already used on page 85, in the proof of Lemma 5.6.

(3) Consider a subsequence of the sequence $\{u_\rho\}$, again denoted $\{u_\rho\}$, and an element $w \in W^{1,2}(0,T;X)$ such that

$$u_\rho \rightharpoonup w \quad \text{in } W^{1,2}(0,T;X) \quad \text{as } \rho \to 0.$$

It follows from the arguments above that w is a solution to problem (5.1)–(5.2) and therefore, since this last problem has a unique solution denoted u, we find that $w = u$. We conclude that the weak limit in $W^{1,2}(0,T;X)$ is independent of the subsequence extracted and, therefore, the whole sequence $\{u_\rho\}$ converges weakly in $W^{1,2}(0,T;X)$ to u, i.e., (5.71) holds. □

Given a family of functionals $\{j_\rho\}$ that satisfies (5.64), it is sometimes difficult to check conditions (5.65)–(5.65), as well as conditions (5.67)–(5.68). Moreover, note that conditions (5.67)–(5.68) do not represent intrinsic conditions on the initial data of the evolutionary variational inequality (5.1)–(5.2), as they are related to the family of functionals $\{j_\rho\}$ that regularize j. For these reasons, in what follows we replace conditions (5.65)–(5.65) with condition (3.17), already used in Chapters 3 and 4. The interest in this last condition is that it is time independent and therefore it can be easily checked. Also, we replace assumptions (5.67)–(5.68) on the initial data with assumption (5.7). As we shall see in the two lemmas below, condition (3.17) is stronger than conditions (5.65)–(5.65) and, combined with assumption (5.7), allows us to obtain a version of Theorem 5.8 in which assumptions (5.65)–(5.68) are avoided.

Lemma 5.9. *Assume* (5.63) *and* (5.64). *Then condition* (3.17) *implies conditions* (5.65) *and* (5.65).

Proof. Let $\rho > 0$, $v \in L^2(0,T;X)$ and assume that (3.17) holds. We use this condition to see that

$$\begin{aligned}\left| \int_0^T j_\rho(v(t))\,dt - \int_0^T j(v(t))\,dt \right| &\le \int_0^T |j_\rho(v(t)) - j(v(t))|\,dt \\ &\le \int_0^T F(\rho)\,dt \le F(\rho)\,T\end{aligned}$$

and therefore, since $F(\rho) \to 0$ as $\rho \to 0$, we deduce that (5.65) holds.

Consider now a sequence $\{v_\rho\} \subset L^2(0,T;X)$ and let $v \in L^2(0,T;X)$ be such that

$$v_\rho \rightharpoonup v \quad \text{in } L^2(0,T;X). \tag{5.83}$$

We have

$$\int_0^T j_\rho(v_\rho(t))\,dt = \int_0^T \big(j_\rho(v_\rho(t)) - j(v_\rho(t))\big)\,dt + \int_0^T j(v_\rho(t))\,dt \quad \forall\,\rho > 0$$

and, since (3.17) implies that

$$\lim_{\rho\to 0} \int_0^T \big(j_\rho(v_\rho(t)) - j(v_\rho(t))\big)\,dt = 0,$$

we deduce that

$$\liminf_{\rho\to 0} \int_0^T j_\rho(v_\rho(t))\,dt = \liminf_{\rho\to 0} \int_0^T j(v_\rho(t))\,dt. \tag{5.84}$$

The convergence (5.83), assumption (5.63), and the lower semicontinuity result in Corollary 2.32 yield

$$\liminf_{\rho\to 0} \int_0^T j(v_\rho(t))\,dt \geq \int_0^T j(v(t))\,dt. \tag{5.85}$$

We use now (5.84) and (5.85) to see that (5.65) holds. □

Lemma 5.10. *Assume* (5.3), (5.5)–(5.7). *Then condition* (3.17) *implies conditions* (5.67) *and* (5.68).

Proof. Let $\rho > 0$ and assume that (3.17) holds. We use (5.7) and (3.17)(a) to see that

$$a(u_0, v) + j_\rho(v) - (f(0), v)_X \geq j_\rho(v) - j(v) - \delta_0 \geq -\delta_0 - F(\rho) \quad \forall\, v \in X$$

and therefore condition (5.67) holds with $\delta_{0\rho} = \delta_0 + F(\rho)$.

Next, it follows from (3.17)(b) that there exists $\rho^* > 0$ such that $F(\rho) < 1$ for all $\rho \in (0, \rho^*)$. We conclude that $\delta_{0\rho} < \delta_0 + 1$ for all $\rho \in (0, \rho^*)$ and therefore condition (5.68) holds with $M^* = \delta_0 + 1$. □

We now use Lemmas 5.9 and 5.10 to obtain the following version of Theorem 5.8.

Theorem 5.11. *Let X be a Hilbert space and assume that* (5.3), (5.5)–(5.7), (5.63), (5.64), *and* (3.17) *hold. Then:*

(1) For each $\rho > 0$ there exists a unique solution to problem (5.2)–(5.70) *and it satisfies $u_\rho \in W^{1,2}(0, T; X)$.*

(2) There exists a unique solution to problem (5.1)–(5.2) *and it satisfies $u \in W^{1,2}(0, T; X)$.*

(3) The solution u_ρ of problem (5.2)–(5.70) *converges uniformly to the solution u of problem* (5.1)–(5.2), *i.e.,*

$$u_\rho \to u \quad \text{in } C([0,T], X) \quad \text{as } \rho \to 0. \tag{5.86}$$

Proof. The first two parts of the theorem are direct consequences of Lemmas 5.9 and 5.10 combined with Theorem 5.8(1), (2). Therefore, we have to prove only the convergence result (5.86). To this end, consider $\rho > 0$ and choose $v = \dot{u}_\rho(t)$ in (5.1) and $v = \dot{u}(t)$ in (5.2); we add the resulting inequalities to find

$$\begin{aligned} &a(u_\rho(t) - u(t), \dot{u}_\rho(t) - \dot{u}(t)) \\ &\quad \le j(\dot{u}_\rho(t)) - j_\rho(\dot{u}_\rho(t)) + j_\rho(\dot{u}(t)) - j(\dot{u}(t)) \qquad \text{a.e. } t \in (0,T). \end{aligned}$$

We use now assumption (3.17)(a) to see that

$$j(\dot{u}_\rho(t)) - j_\rho(\dot{u}_\rho(t)) + j_\rho(\dot{u}(t)) - j(\dot{u}(t)) \le 2F(\rho) \qquad \text{a.e. } t \in (0,T)$$

and, therefore, combining the last two inequalities yields

$$a(u_\rho(t) - u(t), \dot{u}_\rho(t) - \dot{u}(t)) \le 2F(\rho) \qquad \text{a.e. } t \in (0,T). \tag{5.87}$$

Let $s \in [0,T]$. We integrate (5.87) on $[0,s]$ with the initial conditions $u_\rho(0) = u_0$, $u(0) = u_0$, then we use Proposition 2.30 and (5.3)(b); as a result, we find

$$\|u_\rho(s) - u(s)\|_X^2 \le \frac{4T}{m}\, F(\rho)$$

which, combined with (3.17)(b), implies (5.86). □

The interest in Theorems 5.8 and 5.11 is twofold; first, these theorems provide existence and uniqueness results in the study of the evolutionary variational inequality (5.1)–(5.2) in the case when j is not necessary a Gâteaux differentiable function; secondly, they show that the solution u of problem (5.1)–(5.2) can be obtained as limit of the solutions u_ρ of problem (5.2)–(5.70), in the sense of (5.71) and (5.86), respectively.

The assumptions made in Theorems 5.8 and 5.11 require the function j to be approached by a sequence of functionals, $\{j_\rho\}$, which are more regular, since they are Gâteaux differentiable. For this reason, the method used in the proof of Theorems 5.8 and 5.11 is called the *regularization method*. Its main feature consists in proving the existence of the solution of an "irregular" variational inequality as limit of the solutions of "regular" variational inequalities.

We end this section with the remark that the existence of a unique solution to the Cauchy problem (5.1)–(5.2) can be obtained under weaker assumptions that those in Theorems 5.1, 5.8, and 5.11. Indeed, the following existence and uniqueness result represents a direct version of a result proved in [16, p. 117].

Theorem 5.12. *Let X be a Hilbert space and assume that* (5.3), (5.5)–(5.7), (4.5) *hold. Then there exists a unique solution of the evolutionary variational inequality* (5.1)–(5.2) *and it satisfies* $u \in W^{1,2}(0,T;X)$.

Applying Theorem 5.12 to the case when j is the indicator function of a nonempty, convex, and closed subset $K \subset X$ leads to the following result.

Corollary 5.13. *Let X be a Hilbert space and let $K \subset X$ be a nonempty, convex, and closed subset. Assume that* (5.3), (5.5), (5.6) *hold and, moreover, there exists $\delta_0 \geq 0$ such that $a(u_0, v) \geq (f(0), v)_X - \delta_0$ for all $v \in K$. Then there exists a unique function $u \in W^{1,2}(0,T;X)$ such that $u(0) = u_0$ and*

$$\dot{u}(t) \in K, \quad a(u(t), v - \dot{u}(t)) \geq (f(t), v - \dot{u}(t))_X \quad \forall\, v \in K, \quad a.e.\ t \in (0,T).$$

The proof of Theorem 5.12 requires arguments and preliminary results from the theory of maximal monotone operators that are not included in this book and, therefore, is omitted.

5.3 A Convergence Result

In this section, we study the behavior of the solution of the evolutionary variational inequality (4.1)–(4.2) as the viscosity vanishes, and we prove that it converges to the solution of the evolutionary variational inequality (5.1)–(5.2). We assume in what follows that (5.3), (5.5)–(5.7), and (4.5) hold and, therefore, we deduce from Theorem 5.12 that problem (5.1)–(5.2) has a unique solution $u \in W^{1,2}(0,T;X)$.

Let Θ be a given set of parameters and, for every $\theta \in \Theta$, consider a bilinear form b_θ that satisfies condition (4.4), that is,

$$\left.\begin{array}{l} b_\theta : X \times X \to \mathbb{R} \text{ is a bilinear symmetric form and} \\ \text{(a) there exists } M'(\theta) > 0 \text{ such that} \\ \qquad |b_\theta(u,v)| \leq M'(\theta) \|u\|_X \|v\|_X \quad \forall\, u,\, v \in X. \\ \text{(b) there exists } m'(\theta) > 0 \text{ such that} \\ \qquad b_\theta(v,v) \geq m'(\theta) \|v\|_X^2 \quad \forall\, v \in X. \end{array}\right\} \tag{5.88}$$

Also, for every $\theta \in \Theta$, consider the problem of finding a function $u_\theta : [0,T] \to X$ such that

$$\begin{aligned} & a(u_\theta(t), v - \dot{u}_\theta(t)) + b_\theta(\dot{u}_\theta(t),\, v - \dot{u}_\theta(t)) + j(v) - j(\dot{u}_\theta(t)) \\ & \quad \geq (f(t), v - \dot{u}_\theta(t))_X \quad \forall\, v \in X,\ t \in [0,T], \end{aligned} \tag{5.89}$$

$$u_\theta(0) = u_0. \tag{5.90}$$

We deduce from Theorems 4.1 and 4.5 that problem (5.89)–(5.90) has a unique solution $u_\theta \in W^{2,2}(0,T;X)$.

The following result provides a condition that ensures the convergence of the solution u_θ of problem (5.89)–(5.90) to the solution u of problem (5.1)–(5.2).

Theorem 5.14. *Under the previous assumptions,*

$$\frac{M'(\theta)^2}{m'(\theta)} \to 0 \implies \|u_\theta - u\|_{C([0,T];X)} \to 0. \tag{5.91}$$

The convergence (5.91) is understood in the sense described on page 48, i.e., for every sequence $\{\theta_n\} \subset \Theta$ such that $\frac{M'(\theta_n)^2}{m'(\theta_n)} \to 0$ as $n \to \infty$, one has $\|u_{\theta_n} - u\|_{C([0,T];X)}$ as $n \to \infty$.

Proof. The equalities and inequalities below hold for almost every $t \in (0,T)$. We choose $v = \dot{u}(t)$ in (5.89), $v = \dot{u}_\theta(t)$ in (5.1), and add the resulting inequalities to obtain

$$a(u_\theta(t) - u(t), \dot{u}(t) - \dot{u}_\theta(t)) + b_\theta(\dot{u}_\theta(t), \dot{u}(t) - \dot{u}_\theta(t)) \geq 0.$$

This implies that

$$\begin{aligned} &a(u_\theta(t) - u(t), \dot{u}_\theta(t) - \dot{u}(t)) + b_\theta(\dot{u}_\theta(t) - \dot{u}(t), \dot{u}_\theta(t) - \dot{u}(t)) \\ &\quad \leq b_\theta(\dot{u}(t), \dot{u}(t) - \dot{u}_\theta(t)) \end{aligned}$$

and, using assumption (5.88), yields

$$\begin{aligned} &a(u_\theta(t) - u(t), \dot{u}_\theta(t) - \dot{u}(t)) + m'(\theta)\|\dot{u}_\theta(t) - \dot{u}(t)\|_X^2 \\ &\quad \leq M'(\theta)\, \|\dot{u}(t)\|_X \|\dot{u}_\theta(t) - \dot{u}(t)\|_X. \end{aligned} \tag{5.92}$$

We now combine (5.92) with the elementary inequality

$$\|\dot{u}(t)\|_X \, \|\dot{u}_\theta(t) - \dot{u}(t)\|_X \leq \frac{M'(\theta)}{4m'(\theta)} \, \|\dot{u}(t)\|_X^2 + \frac{m'(\theta)}{M'(\theta)} \, \|\dot{u}_\theta(t) - \dot{u}(t)\|_X^2$$

and, as a result, we obtain

$$a(u_\theta(t) - u(t), \dot{u}_\theta(t) - \dot{u}(t)) \leq \frac{M'(\theta)^2}{4m'(\theta)} \, \|\dot{u}(t)\|_X^2.$$

Let $s \in [0,T]$. We integrate the previous inequality on $[0,s]$ with the initial conditions $u_\theta(0) = u(0) = u_0$ and use assumption (5.3) to obtain

$$\frac{m}{2} \, \|u_\theta(s) - u(s)\|_X^2 \leq \frac{M'(\theta)^2}{4m'(\theta)} \int_0^s \|\dot{u}(t)\|_X^2 \, dt.$$

This last inequality shows that

$$\|u_\theta(s) - u(s)\|_X \leq \sqrt{\frac{M'(\theta)^2}{2mm'(\theta)}} \, \|\dot{u}\|_{L^2(0,T;X)}. \tag{5.93}$$

Assume now that

$$\frac{M'(\theta)^2}{m'(\theta)} \to 0. \tag{5.94}$$

Then, it follows from estimate (5.93) that $\|u_\theta - u\|_{C([0,T];X)} \to 0$, which completes the proof. □

We conclude from Theorem 5.14 that we can approach the solution of the inviscid evolutionary variational inequality (5.1)–(5.2) with the solution of the evolutionary variational inequality with viscosity (5.89)–(5.90) if the ratio $\frac{M'(\theta)^2}{m'(\theta)}$ is small enough. Note that we already used a similar result in the proof of Theorem 5.1. Indeed, there we have considered the regularized problems (5.9)–(5.10), which are of the form (5.89)–(5.90) with $\theta = \frac{1}{n}$ and $b_\theta(u,v) = \frac{1}{n}(u,v)_X$. It is clear that in this case (5.94) holds as $n \to \infty$ and, therefore, by Theorem 5.14 we recover the convergence result presented in Lemma 5.5.

Finally, a brief comparison between the results presented in Section 4.1 in the study of evolutionary variational inequality with viscosity (Theorems 4.1 and 4.5) and the results presented in Sections 5.1 and 5.2 in the study of inviscid evolutionary variational inequalities (Theorems 5.1, 5.8, 5.11 and 5.12) leads to the following remarks:

(a) Under the same regularity (say $f \in W^{1,2}(0,T;X)$ and $u_0 \in X$) on the data, the regularity of the solution of problem (4.1)–(4.2) is higher than the regularity of the solution of problem (5.1)–(5.2) ($W^{2,2}(0,T;X)$ instead of $W^{1,2}(0,T;X)$, respectively);

(b) To solve problem (5.1)–(5.2), one needs a compatibility assumption on the initial data ((5.7) or (5.67), (5.68)); in contrast, to solve problem (4.1)–(4.2), such an assumption is not needed.

The two remarks above lead to the conclusion that adding viscosity has a *smoothing effect* on the solution of evolutionary variational inequalities. This feature combined with the convergence result (5.91) justifies the use of numerical methods based on regularization with viscosity in the study of (5.1)–(5.2).

5.4 Evolutionary Quasivariational Inequalities

In this section, we study an evolutionary version of the quasivariational inequality (3.29). To this end, we assume in what follows that X is a separable Hilbert space and let $a : X \times X \to \mathbb{R}$, $j : X \times X \to \mathbb{R}$, $f : [0,T] \to X$ be given. We consider the problem of finding a function $u : [0,T] \to X$ such that

$$\begin{aligned} &a(u(t), v - \dot{u}(t)) + j(u(t), v) - j(u(t), \dot{u}(t)) \\ &\quad \geq (f(t), v - \dot{u}(t))_X \quad \forall\, v \in X, \quad \text{a.e. } t \in (0,T), \end{aligned} \tag{5.95}$$

$$u(0) = u_0. \tag{5.96}$$

Here u_0 represents a given initial value, and we need to prescribe an initial condition since the problem is evolutionary.

In the study of (5.95)–(5.96), we assume that a is a bilinear symmetric continuous and X-elliptic form, i.e., it satisfies (5.3); moreover, the data f and u_0 satisfy:

$$f \in W^{1,\infty}(0,T;X). \tag{5.97}$$

$$u_0 \in X. \tag{5.98}$$

$$a(u_0,v) + j(u_0,v) \geq (f(0),v)_X \quad \forall\, v \in X. \tag{5.99}$$

Note that assumption (5.99) represents a compatibility assumption of the form (5.7), with $\delta_0 = 0$.

We consider now the following assumptions on the functional j.

$$\text{For all } \eta \in X,\ j(\eta,\,\cdot) \text{ is a seminorm on } X. \tag{5.100}$$

$$\left.\begin{array}{l}\text{For all sequences } \{\eta_n\} \subset X \text{ and } \{u_n\} \subset X \text{ such that}\\ \eta_n \rightharpoonup \eta \in X,\ u_n \rightharpoonup u \in X \text{ and for every } v \in X,\\ \text{the inequality below holds:}\\ \limsup\limits_{n\to\infty}\, [j(\eta_n,v) - j(\eta_n,u_n)] \leq j(\eta,v) - j(\eta,u).\end{array}\right\} \tag{5.101}$$

$$\left.\begin{array}{l}\text{There exists } \alpha \in (0,m) \text{ such that}\\ j(u,v-u) - j(v,v-u) \leq \alpha\|u-v\|_X^2 \quad \forall\, u,v \in X.\end{array}\right\} \tag{5.102}$$

$$\left.\begin{array}{l}\text{There exist two functions } a_1 : X \to \mathbb{R} \text{ and } a_2 : X \to \mathbb{R}\\ \text{that map bounded sets in } X \text{ into bounded sets in } \mathbb{R}\\ \text{such that } \ j(\eta,u) \leq a_1(\eta)\,\|u\|_X^2 + a_2(\eta) \quad \forall\, \eta,u \in X,\\ \text{and } a_1(0_X) < m - \alpha.\end{array}\right\} \tag{5.103}$$

$$\left.\begin{array}{l}\text{For every sequence } \{\eta_n\} \subset X \text{ with } \eta_n \rightharpoonup \eta \in X,\\ \text{and every bounded sequence } \{u_n\} \subset X, \text{ one has}\\ \lim\limits_{n\to\infty}\, [j(\eta_n,u_n) - j(\eta,u_n)] = 0.\end{array}\right\} \tag{5.104}$$

$$\left.\begin{array}{l}\text{For every } s \in (0,T] \text{ and every function } u,v \in W^{1,\infty}(0,T;X)\\ \text{with } u(0) = v(0),\ u(s) \neq v(s), \text{ the inequality below holds:}\\ \displaystyle\int_0^s [j(u(t),\dot v(t)) - j(u(t),\dot u(t)) + j(v(t),\dot u(t)) - j(v(t),\dot v(t))]dt\\ < \dfrac{m}{2}\,\|u(s)-v(s)\|_X^2.\end{array}\right\} \tag{5.105}$$

$$\left.\begin{array}{l}\text{There exists } \beta \in (0,\frac{m}{2}) \text{ such that for every } s \in (0,T] \text{ and}\\ \text{every function } u,v \in W^{1,\infty}(0,T;X) \text{ with } u(s) \neq v(s),\\ \text{the inequality below holds:}\\ \displaystyle\int_0^s [j(u(t),\dot v(t)) - j(u(t),\dot u(t)) + j(v(t),\dot u(t)) - j(v(t),\dot v(t))]dt\\ \leq \beta\,\|u(s)-v(s)\|_X^2.\end{array}\right\} \tag{5.106}$$

Note that assumptions (5.100) and (5.101) have already been used in Section 3.4 in the study of elliptic quasivariational inequalities, see (3.36) and (3.37), respectively.

Our main result in the study of problem (5.95)–(5.96) is the following.

Theorem 5.15. *Let X be a separable Hilbert space and assume that* (5.3) *and* (5.97)–(5.99) *hold. Then:*

(1) Under the assumptions (5.100)–(5.104), *there exists at least one solution $u \in W^{1,\infty}(0,T;X)$ to problem* (5.95)–(5.96).

(2) Under the assumptions (5.100)–(5.105), *there exists a unique solution $u \in W^{1,\infty}(0,T;X)$ to problem* (5.95)–(5.96).

(3) Under the assumptions (5.100)–(5.104) *and* (5.106), *there exists a unique solution $u = u(f, u_0) \in W^{1,\infty}(0,T;X)$ to problem* (5.95)–(5.96), *and the mapping $(f, u_0) \mapsto u$ is Lipschitz continuous from $W^{1,\infty}(0,T;X) \times X$ to $C([0,T];X)$.*

The proof of Theorem 5.15 will be carried out in several steps by using a time discretization method, compactness and lower semicontinuity arguments. We suppose in what follows that (5.3), (5.97)–(5.104) hold. Let $n \in \mathbb{N}$. We consider the following implicit scheme: find $u_n^{i+1} \in X$ such that

$$a\left(u_n^{i+1}, v - \frac{n}{T}(u_n^{i+1} - u_n^i)\right) + j(u_n^{i+1}, v) - j\left(u_n^{i+1}, \frac{n}{T}(u_n^{i+1} - u_n^i)\right)$$
$$\geq \left(f\left(\frac{T(i+1)}{n}\right), v - \frac{n}{T}(u_n^{i+1} - u_n^i)\right)_X \quad \forall\, v \in X, \tag{5.107}$$

where $u_n^0 = u_0$, $i = 0, 1, \ldots, n-1$.

In the first step, we prove the solvability of the elliptic quasivariational inequality (5.107) and we provide estimates on the solution to this problem.

Lemma 5.16. *There exists at least one solution u_n^{i+1} to the quasivariational inequality* (5.107), *for $i = 0, 1, \ldots, n-1$. Moreover, the solution satisfies*

$$\|u_n^{i+1}\|_X^2 \leq \frac{1}{m - \alpha - a_1(0_X)}\left(\left\|f\left(\frac{T(i+1)}{n}\right)\right\|_X \|u_n^{i+1}\|_X + a_2(0_X)\right), \tag{5.108}$$

$$\|u_n^{i+1} - u_n^i\|_X \leq \frac{1}{m - \alpha}\left\|f\left(\frac{T(i+1)}{n}\right) - f\left(\frac{Ti}{n}\right)\right\|_X, \tag{5.109}$$

for all $i = 0, 1, \ldots, n-1$.

Proof. Let $i \in \{0, 1, \dots, n-1\}$. We set $w = \frac{T}{n} v + u_n^i$ and use the properties of a and j to see that (5.107) is equivalent to the inequality

$$a(u_n^{i+1}, w - u_n^{i+1}) + j(u_n^{i+1}, w - u_n^i) - j(u_n^{i+1}, u_n^{i+1} - u_n^i) \geq \left(f\left(\frac{T(i+1)}{n} \right), w - u_n^{i+1} \right)_X \quad \forall\, w \in X. \tag{5.110}$$

Using Theorem 3.10(1), we obtain the existence of a solution u_n^{i+1} to (5.110) and, therefore, the equivalence of problems (5.107) and (5.110) yields the first part of the lemma.

Choosing now $w = 0_X$ in (5.110) and using (5.100), we find

$$a(u_n^{i+1}, u_n^{i+1}) \leq \left(f\left(\frac{T(i+1)}{n} \right), u_n^{i+1} \right)_X + j(u_n^{i+1}, -u_n^i) - j(u_n^{i+1}, u_n^{i+1} - u_n^i)$$
$$\leq \left\| f\left(\frac{T(i+1)}{n} \right) \right\|_X \|u_n^{i+1}\|_X + j(u_n^{i+1}, -u_n^{i+1}),$$

and using (5.3)(b) yields

$$m \, \|u_n^{i+1}\|_X^2 \leq \left\| f\left(\frac{T(i+1)}{n} \right) \right\|_X \|u_n^{i+1}\|_X + j(u_n^{i+1}, -u_n^{i+1}). \tag{5.111}$$

Choosing $u = u_n^{i+1}$ and $v = 0_X$ in (5.102) and using (5.103) with $\eta = 0_X$, we obtain

$$j(u_n^{i+1}, -u_n^{i+1}) \leq \alpha \, \|u_n^{i+1}\|_X^2 + j(0_X, -u_n^{i+1}) \leq (\alpha + a_1(0_X)) \, \|u_n^{i+1}\|_X^2 + a_2(0_X). \tag{5.112}$$

Estimate (5.108) results from (5.111) and (5.112), since $a_1(0_X) < m - \alpha$.

Using (5.100), it follows that

$$j(u, 0_X) = 0 \quad \forall\, u \in X \tag{5.113}$$

and therefore, by setting $w = u_n^i$ in (5.110) and using (5.113), we obtain

$$a(u_n^{i+1}, u_n^{i+1} - u_n^i) \leq \left(f\left(\frac{T(i+1)}{n} \right), u_n^{i+1} - u_n^i \right)_X - j(u_n^{i+1}, u_n^{i+1} - u_n^i) \quad \forall\, i = 0, 1, \dots, n-1. \tag{5.114}$$

Using again (5.110) with $i-1$ in place of i and $w = u_n^{i+1}$, we find

$$a(u_n^i, u_n^{i+1} - u_n^i) + j(u_n^i, u_n^{i+1} - u_n^{i-1}) - j(u_n^i, u_n^i - u_n^{i-1}) \geq \left(f\left(\frac{Ti}{n} \right), u_n^{i+1} - u_n^i \right)_X \quad \forall\, i = 1, \dots, n-1$$

and, using (5.100), we obtain

$$-a(u_n^i, u_n^{i+1} - u_n^i) \le -\left(f\left(\frac{Ti}{n}\right), u_n^{i+1} - u_n^i\right)_X + j(u_n^i, u_n^{i+1} - u_n^i) \quad \forall\, i = 1, \ldots, n-1.$$

Note that, since $u_n^0 = u_0$, assumption (5.99) yields

$$-a(u_n^0, u_n^1 - u_n^0) \le -(f(0), u_n^1 - u_n^0)_X + j(u_n^0, u_n^1 - u_n^0).$$

We conclude by the last two inequalities that

$$-a(u_n^i, u_n^{i+1} - u_n^i) \le -\left(f\left(\frac{Ti}{n}\right), u_n^{i+1} - u_n^i\right)_X + j(u_n^i, u_n^{i+1} - u_n^i) \quad \forall\, i = 0, 1, \ldots, n-1. \tag{5.115}$$

It follows now from (5.3)(b), (5.114), (5.115), and (5.102) that

$$\begin{aligned} m\,\|u_n^{i+1} - u_n^i\|_X^2 &\le a(u_n^{i+1} - u_n^i, u_n^{i+1} - u_n^i) \\ &\le \left(f\left(\frac{T(i+1)}{n}\right) - f\left(\frac{Ti}{n}\right), u_n^{i+1} - u_n^i\right)_X \\ &\quad - j(u_n^{i+1}, u_n^{i+1} - u_n^i) + j(u_n^i, u_n^{i+1} - u_n^i) \\ &\le \left\| f\left(\frac{T(i+1)}{n}\right) - f\left(\frac{Ti}{n}\right)\right\|_X \|u_n^{i+1} - u_n^i\|_X + \alpha\,\|u_n^{i+1} - u_n^i\|_X^2 \\ &\quad \forall\, i = 0, 1, \ldots, n-1, \end{aligned}$$

which implies (5.109). □

We now consider the functions $u_n : [0,T] \to X$ and $\tilde{u}_n : [0,T] \to X$ defined as follows:

$$u_n(0) = u_0, \quad u_n(t) = u_n^i + \frac{nt - Ti}{T}(u_n^{i+1} - u_n^i) \quad \forall\, t \in \left(\frac{Ti}{n}, \frac{T(i+1)}{n}\right], \tag{5.116}$$

$$\tilde{u}_n(0) = u_0, \quad \tilde{u}_n(t) = u_n^{i+1} \quad \forall\, t \in \left(\frac{Ti}{n}, \frac{T(i+1)}{n}\right], \tag{5.117}$$

where $u_n^0 = u_0$, u_n^{i+1} solves (5.107) and $i = 0, 1, \ldots, n-1$.

In the next step, we provide convergence results involving the sequences $\{u_n\}$ and $\{\tilde{u}_n\}$.

Lemma 5.17. *There exists an element $u \in W^{1,\infty}(0,T;X)$ and subsequences of the sequences $\{u_n\}$ and $\{\tilde{u}_n\}$, again denoted $\{u_n\}$ and $\{\tilde{u}_n\}$, respectively, such that:*

$$u_n \rightharpoonup^* u \quad \text{in } L^\infty(0,T;X), \tag{5.118}$$

$$\dot{u}_n \rightharpoonup^* \dot{u} \quad \text{in } L^\infty(0,T;X), \tag{5.119}$$

$$\tilde{u}_n(t) \rightharpoonup u(t) \quad \text{in } X, \quad \text{a.e. } t \in (0,T). \tag{5.120}$$

Proof. Let $n \in \mathbb{N}$. Using (5.116), it follows that $u_n : [0,T] \to X$ is an absolutely continuous function and its derivative is given by

$$\dot{u}_n(t) = \frac{n}{T}\,(u_n^{i+1} - u_n^i) \quad \text{a.e. } t \in \left(\frac{Ti}{n}, \frac{T(i+1)}{n}\right) \quad \forall\, i = 0, 1, \ldots, n-1. \tag{5.121}$$

Therefore, from (5.116), (5.121), (5.108), and (5.109), we deduce that

$$\|u_n(t)\|_X \le \|u_0\|_X + \frac{1}{m-\alpha}\left\| f\left(\frac{T}{n}\right) - f(0)\right\|_X \quad \text{a.e. } t \in \left(0, \frac{T}{n}\right),$$

$$\|u_n(t)\|_X \le \frac{1}{(m-\alpha-a_1(0_X))^{1/2}}\left(\left\| f\left(\frac{Ti}{n}\right)\right\|_X \|u_n^i\|_X + a_2(0_X)\right)^{1/2}$$
$$+ \frac{1}{m-\alpha}\left\| f\left(\frac{T(i+1)}{n}\right) - f\left(\frac{Ti}{n}\right)\right\|_X$$
$$\text{a.e. } t \in \left(\frac{Ti}{n}, \frac{T(i+1)}{n}\right) \quad \forall\, i = 1, \ldots, n-1,$$

$$\|\dot{u}_n(t)\|_X \le \frac{1}{m-\alpha}\cdot\frac{n}{T}\left\| f\left(\frac{T(i+1)}{n}\right) - f\left(\frac{Ti}{n}\right)\right\|_X$$
$$\text{a.e. } t \in \left(\frac{Ti}{n}, \frac{T(i+1)}{n}\right) \quad \forall\, i = 0, 1, \ldots, n-1.$$

Keeping in mind the regularity (5.97) of f and estimate (5.108), from the previous inequalities it follows that $u_n \in W^{1,\infty}(0,T;X)$ and

$$\|u_n\|_{W^{1,\infty}(0,T;X)} \le c. \tag{5.122}$$

Here and everywhere in this section, c represents a positive constant that may depend on f and u_0 but does not depend on n and whose value may change from place to place.

Inequality (5.122) shows that $\{u_n\}$ is a bounded sequence in the space $W^{1,\infty}(0,T;X)$; therefore, using a standard compactness argument, we deduce that there exists an element $u \in W^{1,\infty}(0,T;X)$ and a subsequence of $\{u_n\}$, again denoted $\{u_n\}$, such that (5.118) and (5.119) hold.

We turn now to the proof of (5.120). To this end, we remark that the convergences (5.118) and (5.119) imply that

$$u_n(t) \rightharpoonup u(t) \quad \text{in } X, \quad \text{for all } t \in [0,T]. \tag{5.123}$$

Moreover, using again (5.116), (5.117), and (5.109), we find

$$\begin{aligned}\|u_n(t) - \tilde{u}_n(t)\|_X &= \left(1 - \frac{nt - Ti}{T}\right)\|u_n^{i+1} - u_n^i\|_X \\ &\le \frac{1}{m-\alpha}\left\|f\left(\frac{T(i+1)}{n}\right) - f\left(\frac{Ti}{n}\right)\right\|_X \\ &\qquad \forall\, t \in \left(\frac{Ti}{n}, \frac{T(i+1)}{n}\right], \quad i = 0, 1, \ldots, n-1\end{aligned}$$

and, keeping in mind the regularity (5.97) of f, we deduce that

$$\|u_n - \tilde{u}_n\|_{L^\infty(0,T;X)} \le \frac{T}{n(m-\alpha)}\,\|\dot{f}\|_{L^\infty(0,T;X)}.$$

This inequality proves that

$$u_n - \tilde{u}_n \to 0 \quad \text{in } L^\infty(0,T;X) \tag{5.124}$$

and, therefore,

$$u_n(t) - \tilde{u}_n(t) \to 0 \quad \text{in } X, \ \text{a.e. } t \in (0,T). \tag{5.125}$$

The convergence (5.120) is now a consequence of (5.123) and (5.125). □

In the next two steps, we prove additional convergence and semicontinuity results. To this end, for every $n \in \mathbb{N}$ consider the function $f_n : [0,T] \to X$ defined as follows:

$$f_n(0) = f(0), \quad f_n(t) = f\left(\frac{T(i+1)}{n}\right) \tag{5.126}$$

$$\forall\, t \in \left(\frac{Ti}{n}, \frac{T(i+1)}{n}\right], \ i = 0, 1, \ldots, n-1.$$

Everywhere in the sequel, u will denote an element of $W^{1,\infty}(0,T;X)$ that satisfies the properties in Lemma 5.17 and $\{u_n\}$, $\{\tilde{u}_n\}$, $\{f_n\}$ will represent appropriate subsequences of the sequences $\{u_n\}$, $\{\tilde{u}_n\}$ and $\{f_n\}$, respectively, which satisfy (5.118)–(5.120).

Lemma 5.18. *The following properties hold:*

$$\lim_{n\to\infty} \int_0^T a(\tilde{u}_n(t), g(t))\, dt = \int_0^T a(u(t), g(t))\, dt \quad \forall\, g \in L^2(0,T;X), \quad (5.127)$$

$$\liminf_{n\to\infty} \int_0^T a(\tilde{u}_n(t), \dot{u}_n(t))\, dt \geq \int_0^T a(u(t), \dot{u}(t))\, dt, \quad (5.128)$$

$$\lim_{n\to\infty} \int_0^T (f_n(t), g(t) - \dot{u}_n(t))_X\, dt$$

$$= \int_0^T (f(t), g(t) - \dot{u}(t))_X\, dt \quad \forall\, g \in L^2(0,T;X). \quad (5.129)$$

Proof. It follows from convergences (5.118) and (5.124) that $\tilde{u}_n \rightharpoonup u$ in $L^2(0,T;X)$ and therefore, by using (5.3), we deduce (5.127). Using again (5.3)(a), (5.124), and (5.122), we find

$$\lim_{n\to\infty} \int_0^T a(\tilde{u}_n(t) - u_n(t), \dot{u}_n(t))\, dt = 0 \quad (5.130)$$

and, from (5.123), $u_n(0) = u_0$, and semicontinuity arguments already used in this chapter, we obtain

$$\liminf_{n\to\infty} \int_0^T a(u_n(t), \dot{u}_n(t))\, dt \geq \int_0^T a(u(t), \dot{u}(t))\, dt. \quad (5.131)$$

The inequality (5.128) is now a consequence of (5.130) and (5.131).

Finally, from (5.97) and (5.126), we obtain that the sequence $\{f_n\}$ converges uniformly to f on $[0,T]$, i.e.,

$$\max_{t\in[0,T]} \|f_n(t) - f(t)\|_X \to 0 \quad \text{as } n \to \infty. \quad (5.132)$$

The convergence (5.129) is now a consequence of (5.119) and (5.132). □

Lemma 5.19. *The following properties hold:*

$$\limsup_{n\to\infty} \int_0^T j(\tilde{u}_n(t), g(t))\, dt \leq \int_0^T j(u(t), g(t))\, dt \quad \forall\, g \in L^2(0,T;X), \quad (5.133)$$

$$\limsup_{n\to\infty} \int_0^T [j(u(t), \dot{u}_n(t)) - j(\tilde{u}_n(t), \dot{u}_n(t))]\, dt \leq 0, \quad (5.134)$$

$$\liminf_{n\to\infty} \int_0^T j(u(t), \dot{u}_n(t))\, dt \geq \int_0^T j(u(t), \dot{u}(t))\, dt. \quad (5.135)$$

Proof. Let $g \in L^2(0,T;X)$. Using (5.117), (5.108), and (5.97), it follows that $\{\tilde{u}_n(t)\}$ is a bounded sequence in X, for all $t \in [0,T]$. Therefore, by assumption (5.103), we deduce that there exists $c > 0$ such that

$$j(\tilde{u}_n(t), g(t)) \le c\left(\|g(t)\|_X^2 + 1\right) \quad \text{a.e. } t \in (0,T), \quad \forall\, n \in \mathbb{N}.$$

This inequality allows us to use Fatou's lemma to obtain

$$\limsup_{n\to\infty} \int_0^T j(\tilde{u}_n(t), g(t))\, dt \le \int_0^T \limsup_{n\to\infty} j(\tilde{u}_n(t), g(t))\, dt. \tag{5.136}$$

We use now (5.120) and assumption (5.104) to find that

$$\lim_{n\to\infty} j(\tilde{u}_n(t), g(t)) = j(u(t), g(t)) \quad \text{a.e. } t \in (0,T). \tag{5.137}$$

Inequality (5.133) is now a consequence of (5.136) and (5.137).

Next, we use again assumption (5.103) and (5.122) to deduce that there exists $c > 0$, which depends on u, such that

$$|j(u(t), \dot{u}_n(t)) - j(\tilde{u}_n(t), \dot{u}_n(t))| \le c \quad \text{a.e. } t \in (0,T), \quad \forall\, n \in \mathbb{N}.$$

This inequality allows us to apply again Fatou's lemma to obtain

$$\begin{aligned} &\limsup_{n\to\infty} \int_0^T [j(u(t), \dot{u}_n(t)) - j(\tilde{u}_n(t), \dot{u}_n(t))] dt \\ &\quad \le \int_0^T \limsup_{n\to\infty} [j(u(t), \dot{u}_n(t)) - j(\tilde{u}_n(t), \dot{u}_n(t))] dt. \end{aligned} \tag{5.138}$$

Moreover, using (5.120), (5.122), and assumption (5.104), we deduce that

$$\lim_{n\to\infty} \left[j(u(t), \dot{u}_n(t)) - j(\tilde{u}_n(t), \dot{u}_n(t))\right] = 0 \quad \text{a.e. } t \in (0,T). \tag{5.139}$$

Inequality (5.134) follows now from (5.138) and (5.139).

Finally, inequality (5.135) follows from standard semicontinuity arguments, keeping in mind (5.100), (5.103), and (5.119). □

We have now all the ingredients to prove Theorem 5.15.

Proof. (1) Using inequality (5.107) and equalities (5.117), (5.121), and (5.126), we obtain

$$\begin{aligned} &a(\tilde{u}_n(t), v - \dot{u}_n(t))_X + j(\tilde{u}_n(t), v) - j(\tilde{u}_n(t), \dot{u}_n(t)) \\ &\quad \ge (f_n(t), v - \dot{u}_n(t))_X \quad \forall\, v \in X, \text{ a.e. } t \in (0,T). \end{aligned}$$

This inequality and assumption (5.103) yield

$$\int_0^T a(\tilde{u}_n(t), g(t) - \dot{u}_n(t))\, dt + \int_0^T j(\tilde{u}_n(t), g(t))\, dt - \int_0^T j(\tilde{u}_n(t), \dot{u}_n(t))\, dt$$

$$\geq \int_0^T (f_n(t), g(t) - \dot{u}_n(t)_X)\, dt \quad \forall\, g \in L^2(0,T;X).$$

Using now (5.127)–(5.129), (5.133)–(5.135), by passing to the lower limit in the previous inequality we find

$$\int_0^T a(u(t), g(t) - \dot{u}(t))\, dt + \int_0^T j(u(t), g(t))\, dt - \int_0^T j(u(t), \dot{u}(t))\, dt$$

$$\geq \int_0^T (f(t), g(t) - \dot{u}(t))_X\, dt \quad \forall\, g \in L^2(0,T;X). \tag{5.140}$$

We now use in (5.140) a classical application of Lebesgue point for L^1 functions, Theorem 2.20, and obtain that $u \in W^{1,\infty}(0,T;X)$ satisfies inequality (5.95); moreover, from (5.116) and (5.123) we deduce (5.96), which concludes the proof of this part of the theorem.

(2) Consider two solutions $u_1, u_2 \in W^{1,\infty}(0,T;X)$ to the Cauchy problem (5.95)–(5.96). The inequalities below hold for all $v \in X$ and a.e. $t \in (0,T)$:

$$a(u_1(t), v - \dot{u}_1(t)) + j(u_1(t), v) - j(u_1(t), \dot{u}_1(t)) \geq (f(t), v - \dot{u}_1(t))_X,$$

$$a(u_2(t), v - \dot{u}_2(t)) + j(u_2(t), v) - j(u_2(t), \dot{u}_2(t)) \geq (f(t), v - \dot{u}_2(t))_X.$$

We take $v = \dot{u}_2(t)$ in the first inequality, $v = \dot{u}_1(t)$ in the second inequality, then we add the resulting inequalities and use (5.3) to obtain

$$\begin{aligned}
&\frac{1}{2}\frac{d}{dt} a(u_1(t) - u_2(t), u_1(t) - u_2(t)) \\
&\quad \leq j(u_1(t), \dot{u}_2(t)) - j(u_1(t), \dot{u}_1(t)) + j(u_2(t), \dot{u}_1(t)) \\
&\qquad - j(u_2(t), \dot{u}_2(t)) \quad \text{a.e. } t \in (0,T).
\end{aligned} \tag{5.141}$$

Moreover, from (5.96) we have

$$u_1(0) = u_2(0) = u_0. \tag{5.142}$$

Arguing by contradiction, let us suppose that $u_1 \neq u_2$. Then there exists $s \in (0,T]$ such that

$$u_1(s) \neq u_2(s). \tag{5.143}$$

Integrating (5.141) over $[0, s]$ by using (5.3) and (5.142) yields

$$\frac{m}{2}\|u_1(s) - u_2(s)\|_X^2$$
$$\leq \int_0^s \big[j(u_1(t), \dot{u}_2(t)) - j(u_1(t), \dot{u}_1(t)) + j(u_2(t), \dot{u}_1(t)) - j(u_2(t), \dot{u}_2(t))\big]\,dt. \tag{5.144}$$

In view of (5.142) and (5.143), inequality (5.144) is in contradiction with assumption (5.105). Thus, we conclude that $u_1 = u_2$, i.e., the solution of problem (5.95)–(5.96) is unique.

(3) The unique solvability of the Cauchy problem (5.95)–(5.96) follows from (2) since assumption (5.106) implies (5.105). Consider now two sets of data $f_i \in W^{1,\infty}(0,T;X)$ and $u_{0i} \in X$ such that (5.99) holds, for $i = 1, 2$. We denote in the sequel by $u_i \in W^{1,\infty}(0,T;X)$ the solution of the Cauchy problem (5.95)–(5.96) for the data f_i and u_{0i}. A computation similar to that used to obtain (5.141) leads to the inequality

$$\frac{1}{2}\frac{d}{dt}a(u_1(t) - u_2(t), u_1(t) - u_2(t))$$
$$\leq j(u_1(t), \dot{u}_2(t)) - j(u_1(t), \dot{u}_1(t)) + j(u_2(t), \dot{u}_1(t)) - j(u_2(t), \dot{u}_2(t))$$
$$+(f_1(t) - f_2(t), \dot{u}_1(t) - \dot{u}_2(t))_X \quad \text{a.e. } t \in (0,T).$$

We suppose in what follows that $s \in (0,T]$ is such that $u_1(s) \neq u_2(s)$. Integrating over $[0, s]$ the previous inequality and using the initial conditions $u_i(0) = u_{0i}$ and (5.3) yields

$$\frac{m}{2}\|u_1(s) - u_2(s)\|_X^2 \leq \frac{M}{2}\|u_{01} - u_{02}\|_X^2$$
$$+ \int_0^s \big[j(u_1(t), \dot{u}_2(t)) - j(u_1(t), \dot{u}_1(t)) + j(u_2(t), \dot{u}_1(t)) - j(u_2(t), \dot{u}_2(t))\big]\,dt$$
$$+ \int_0^s (f_1(t) - f_2(t), \dot{u}_1(t) - \dot{u}_2(t))_X\,dt.$$

In view of the assumption (5.106), we obtain

$$\left(\frac{m}{2} - \beta\right)\|u_1(s) - u_2(s)\|_X^2$$
$$\leq \frac{M}{2}\|u_{01} - u_{02}\|_X^2 + \int_0^s (f_1(t) - f_2(t), \dot{u}_1(t) - \dot{u}_2(t))_X\,dt, \tag{5.145}$$

for some $\beta \in (0, \frac{m}{2})$. Let $\delta \in (0, m - 2\beta)$; using the inequality

$$ab \leq \frac{a^2}{2\delta} + \frac{\delta b^2}{2}$$

we obtain

$$
\begin{aligned}
&\int_0^s (f_1(t) - f_2(t), \dot{u}_1(t) - \dot{u}_2(t))_X \, dt \\
&\quad = (f_1(s) - f_2(s), u_1(s) - u_2(s))_X - (f_1(0) - f_2(0), u_{01} - u_{02})_X \\
&\qquad - \int_0^s (\dot{f}_1(t) - \dot{f}_2(t), u_1(t) - u_2(t))_X \, dt \\
&\quad \le \frac{1}{2\delta} \|f_1(s) - f_2(s)\|_X^2 + \frac{\delta}{2} \|u_1(s) - u_2(s)\|_X^2 \\
&\qquad + \frac{1}{2\delta} \|f_1(0) - f_2(0)\|_X^2 + \frac{\delta}{2} \|u_{01} - u_{02}\|_X^2 \\
&\qquad + \frac{1}{2\delta} \int_0^s \|\dot{f}_1(t) - \dot{f}_2(t)\|_X^2 \, dt + \frac{\delta}{2} \int_0^s \|u_1(t) - u_2(t)\|_X^2 \, dt \\
&\quad \le \frac{c}{2\delta} \|f_1 - f_2\|_{W^{1,\infty}(0,T;X)}^2 + \frac{\delta}{2} \|u_1(s) - u_2(s)\|_X^2 \\
&\qquad + \frac{\delta}{2} \|u_{01} - u_{02}\|_X^2 + \frac{\delta}{2} \int_0^s \|u_1(t) - u_2(t)\|_X^2 \, dt,
\end{aligned}
$$

where $c > 0$ depends on T. We now combine (5.145) and the previous inequality to deduce

$$
\begin{aligned}
\|u_1(s) - u_2(s)\|_X^2 \le c_1 \big(&\|u_{01} - u_{02}\|_X^2 + \|f_1 - f_2\|_{W^{1,\infty}(0,T;X)}^2 \big) \\
&+ c_2 \int_0^s \|u_1(t) - u_2(t)\|_X^2 \, dt
\end{aligned}
\tag{5.146}
$$

where c_1, $c_2 > 0$ depend on M, m, β, δ, and T. We note that inequality (5.146) also holds if $u_1(s) = u_2(s)$ or $s = 0$; therefore, we conclude that it holds for all $s \in [0, T]$. This allows us to apply a Gronwall-type argument in (5.146) and, as a result, we obtain

$$
\|u_1(s) - u_2(s)\|_X^2 \le c \big(\|u_{01} - u_{02}\|_X^2 + \|f_1 - f_2\|_{W^{1,\infty}(0,T;X)}^2 \big)
$$

for all $s \in [0, T]$, which concludes the proof. □

We end this section with the remark that if $\varphi : X \to \mathbb{R}_+$ is a continuous seminorm, then the functional j defined by $j(u, v) = \varphi(v)$ for all $u, v \in X$ satisfies the assumptions (5.100)–(5.106). Therefore, from Theorem 5.15 we deduce the following result.

Corollary 5.20. *Let X be a separable Hilbert space and assume that* (5.3), (5.97), (5.98) *hold. Let also $\varphi : X \to \mathbb{R}_+$ be a continuous seminorm and* *assume that* u_0 *satisfies the condition*

$$a(u_0, v) + \varphi(v) \geq (f(0), v)_X \quad \forall\, v \in X. \tag{5.147}$$

Then there exists a unique function $u \in W^{1,\infty}(0,T;X)$ *such that*

$$a(u(t), v - \dot{u}(t)) + \varphi(v) - \varphi(\dot{u}(t)) \geq (f(t), v - \dot{u}(t))_X \quad \forall\, v \in X,\ a.e.\ t \in (0,T), \tag{5.148}$$

$$u(0) = u_0. \tag{5.149}$$

Moreover, the mapping $(f, u_0) \mapsto u$ *is Lipschitz continuous from* $W^{1,\infty}(0,T;X) \times X$ *to* $C([0,T];X)$.

Corollary 5.20 provides an existence and uniqueness result for evolutionary variational inequalities of the form (5.1)–(5.2) studied in Sections 5.1 and 5.2. A direct proof of this result, based again on a time-discretization method, can be found in [60, p. 69–75]. Besides the fact that Corollary 5.20 and Theorems 5.1, 5.8, 5.11, and 5.12 were proved by using different functional arguments, note that they are complementary, since the assumptions on the functionals φ and j that appear in (5.148) and (5.1) are different. We shall use Theorem 5.15 and Corollary 5.20 in Sections 9.4 and 9.3, respectively, in the study of quasistatic antiplane frictional contact problems with elastic materials.

Chapter 6
Volterra-type Variational Inequalities

The variational inequalities studied in this chapter involve the Volterra operator, (4.42), and therefore are history dependent. The term containing this operator appears as a perturbation of the bilinear form a and it is not involved in the function j; even if various other cases may be considered, we made this choice since it is suggested by the structure of variational models that describe the frictional contact of viscoelastic materials with long memory. We consider two types of variational inequalities with a Volterra integral term: the first one is elliptic, and the second one is evolutionary. For both inequalities, we provide existence and uniqueness results. Finally, we study the behavior of the solution with respect to the integral term and with respect to the nondifferentiable function and provide convergence results. The results presented in this chapter will be applied in the study of antiplane frictional contact problems involving viscoelastic materials with long memory. Everywhere in this chapter, X is a real Hilbert space with the inner product $(\cdot, \cdot)_X$, and the norm $\|\cdot\|_X$, and $[0,T]$ denotes the time interval of interest, where $T > 0$.

6.1 Volterra-type Elliptic Variational Inequalities

In this section, we consider the problem of finding $u : [0,T] \to X$ such that

$$
\begin{aligned}
&a(u(t), v - u(t)) + \left(\int_0^t A(t-s)\, u(s)\, ds, v - u(t)\right)_X + j(v) - j(u(t)) \\
&\quad \geq (f(t), v - u(t))_X \quad \forall\, v \in X,\ t \in [0,T], \qquad (6.1)
\end{aligned}
$$

where a is a bilinear form on X, $j : X \to (-\infty, \infty]$, $A : [0,T] \to \mathcal{L}(X)$ and $f : [0,T] \to X$. Note that problem (6.1) does not involve the time derivative of the unknown u and, therefore, it represents an elliptic variational inequality; moreover, since this problem involves the Volterra operator (4.42) with $Y = X$ and $y_0 = 0_X$, we refer to it as a *Volterra-type elliptic variational inequality.*

In the study of (6.1) we assume that:

$$\left.\begin{array}{l} a : X \times X \to \mathbb{R} \text{ is a bilinear symmetric form and} \\ \quad \text{(a) there exists } M > 0 \text{ such that} \\ \qquad |a(u,v)| \le M \|u\|_X \, \|v\|_X \quad \forall\, u,\, v \in X. \\ \quad \text{(b) there exists } m > 0 \text{ such that} \\ \qquad a(v,v) \ge m \, \|v\|_X^2 \quad \forall\, v \in X. \end{array}\right\} \tag{6.2}$$

$$A \in C([0,T]; \mathcal{L}(X)). \tag{6.3}$$

$$j : X \to (-\infty, \infty] \text{ is a proper convex l.s.c. function.} \tag{6.4}$$

$$f \in C([0,T]; X). \tag{6.5}$$

The main result of this section is the following.

Theorem 6.1. *Let X be a Hilbert space and assume that (6.2)–(6.5) hold. Then there exists a unique solution u to problem (6.1). Moreover, the solution satisfies $u \in C([0,T]; X)$.*

Proof. The proof will be carried out in several steps, and it is based on the results on time-dependent elliptic variational inequalities presented in Section 3.4 and a fixed point argument. The steps of the proof are the following.

(i) Let $\eta \in C([0,T]; X)$. We use Theorem 3.12 to see that there exists a unique function $u_\eta \in C([0,T]; X)$ such that

$$\begin{aligned} & a(u_\eta(t), v - u_\eta(t)) + (\eta(t), v - u_\eta(t))_X + j(v) - j(u_\eta(t)) \\ & \quad \ge (f(t), v - u_\eta(t))_X \quad \forall\, v \in X,\ t \in [0,T]. \end{aligned} \tag{6.6}$$

(ii) Next, we consider the operator $\Lambda : C([0,T]; X) \to C([0,T]; X)$ given by

$$\Lambda\eta(t) = \int_0^t A(t-s)\, u_\eta(s)\, ds \quad \forall\, \eta \in C([0,T]; X),\ t \in [0,T] \tag{6.7}$$

and we note that by condition (6.3), the integral in (6.7) is well-defined and the operator Λ takes values in the space $C([0,T]; X)$.

We prove now that Λ has a unique fixed point $\eta^* \in C([0,T]; X)$. To this end, consider two elements $\eta_1, \eta_2 \in C([0,T]; X)$, denote $u_1 = u_{\eta_1}$, $u_2 = u_{\eta_2}$, and let $t \in [0,T]$. Using (6.7), we find that

$$\|\Lambda\eta_1(t) - \Lambda\eta_2(t)\|_X \le \int_0^t \|A(t-s)\|_{\mathcal{L}(X)} \|u_1(s) - u_2(s)\|_X \, ds$$

and, keeping in mind (6.3), it follows that

$$\|\Lambda\eta_1(t) - \Lambda\eta_2(t)\|_X \le c \int_0^t \|u_1(s) - u_2(s)\|_X \, ds \tag{6.8}$$

where $c = \|A\|_{C([0,T];\mathcal{L}(X))}$. Moreover, by using (6.6), after some algebraic manipulations we find that, for all $s \in [0,T]$, the following inequality holds:

$$a(u_1(s) - u_2(s), u_1(s) - u_2(s)) \leq (\eta_1(s) - \eta_2(s), u_2(s) - u_1(s))_X.$$

Therefore, using the X-ellipticity of a, (6.2)(b), we deduce that

$$\|u_1(s) - u_2(s)\|_X \leq \frac{1}{m}\, \|\eta_1(s) - \eta_2(s)\|_X \quad \forall\, s \in [0,T]. \tag{6.9}$$

We combine (6.8) and (6.9) to see that

$$\|\Lambda\eta_1(t) - \Lambda\eta_2(t)\|_X \leq \frac{c}{m} \int_0^t \|\eta_1(s) - \eta_2(s)\|_X\, ds$$

and then use Lemma 1.42 to conclude this step.

(iii) Existence. Let $\eta^* \in C([0,T];X)$ be the fixed point of Λ and let u_{η^*} be the function defined by (6.6) for $\eta = \eta^*$. Since $\Lambda\eta^* = \eta^*$, it follows by (6.6), (6.7) that u_{η^*} satisfies (6.1) and, moreover, $u_{\eta^*} \in C([0,T];X)$.

(iv) Uniqueness. The uniqueness is a consequence of the uniqueness of the fixed point of the operator Λ. A direct proof can be obtained as follows: let u_1, $u_2 \in C([0,T];X)$ be two functions that satisfy (6.1) and let $t \in [0,T]$. Then, after some algebra it follows that

$$a(u_1(t) - u_2(t), u_1(t) - u_2(t)) \leq \left(\int_0^t A(t-s)(u_1(s) - u_2(s))ds, u_2(t) - u_1(t) \right)_X$$

and using assumptions (6.2) and (6.3) yields

$$\|u_1(t) - u_2(t)\|_X \leq \frac{c}{m} \int_0^t \|u_1(s) - u_2(s)\|_X\, ds \tag{6.10}$$

where, again, $c = \|A\|_{C([0,T];\mathcal{L}(X))}$. It follows from (6.10) and Lemma 1.44 that $u_1(t) = u_2(t)$, which concludes the proof. □

Applying Theorem 6.1 to the case when j is the indicator function of a nonempty, convex, and closed subset $K \subset X$ leads to the following result.

Corollary 6.2. *Let X be a Hilbert space, let $K \subset X$ be a nonempty, convex, and closed subset, and assume that* (6.2), (6.3), (6.5) *hold. Then there exists a unique function $u \in C([0,T];X)$ such that*

$$\begin{aligned} u(t) \in K, \quad & a(u(t), v - u(t)) + \left(\int_0^t A(t-s)\, u(s)\, ds, v - u(t) \right)_X \\ & \geq (f(t), v - u(t))_X \quad \forall\, v \in K,\ t \in [0,T]. \end{aligned}$$

We end this section with the following regularity result.

Theorem 6.3. *Under the conditions stated in Theorem 6.1, if there exists* $p \in [1, \infty]$ *such that* $f \in W^{1,p}(0,T;X)$ *and* $A \in W^{1,p}(0,T;\mathcal{L}(X))$, *then* $u \in W^{1,p}(0,T;X)$.

Proof. Denote

$$\eta(t) = \int_0^t A(t-s)u(s)\,ds \quad \forall\, t \in [0,T]. \tag{6.11}$$

It follows that η is differentiable almost everywhere on $(0,T)$ and, moreover,

$$\dot{\eta}(t) = A(0)\,u(t) + \int_0^t \dot{A}(t-s)u(s)\,ds \qquad \text{a.e. } t \in (0,T).$$

Now, since $A \in W^{1,p}(0,T;\mathcal{L}(X))$, we deduce that $\dot{\eta} \in L^p(0,T;X)$, i.e., $\eta \in W^{1,p}(0,T;X)$. We use now (6.1) and (6.11) to see that

$$\begin{aligned} &a(u(t),\, v - u(t)) + j(v) - j(u(t)) \\ &\quad \geq (f(t) - \eta(t),\, v - u(t))_X \quad \forall\, v \in X,\, t \in [0,T]. \end{aligned}$$

Since $f - \eta \in W^{1,p}(0,T;X)$, it follows from Theorem 3.12 that $u \in W^{1,p}(0,T;X)$, which concludes the proof. □

6.2 Volterra-type Evolutionary Variational Inequalities

In this section, we study the evolutionary version of inequality (6.1), which can be formulated as follows: find $u : [0,T] \to X$ such that

$$\begin{aligned} &a(u(t), v - \dot{u}(t)) + \left(\int_0^t A(t-s)\,u(s)\,ds, v - \dot{u}(t) \right)_X + j(v) - j(\dot{u}(t)) \\ &\quad \geq (f(t), v - \dot{u}(t))_X \quad \forall\, v \in X, \text{ a.e. } t \in (0,T), \end{aligned} \tag{6.12}$$

$$u(0) = u_0. \tag{6.13}$$

In the study of this problem, we assume that the bilinear form a satisfies condition (6.2), the function j satisfies condition (6.4), and, moreover,

$$A \in W^{1,2}(0,T;\mathcal{L}(X)), \tag{6.14}$$

$$f \in W^{1,2}(0,T;X), \tag{6.15}$$

$$u_0 \in X, \tag{6.16}$$

$$\left. \begin{array}{l} \text{There exists } \delta_0 \geq 0 \text{ such that} \\ a(u_0, v) + j(v) \geq (f(0), v)_X - \delta_0 \quad \forall\, v \in X. \end{array} \right\} \tag{6.17}$$

Note that condition (6.17) represents a compatibility condition for the initial data of problem (6.12)–(6.13), identical to condition (5.7) used in Chapter 5, in the study of evolutionary variational inequalities. The use of this condition in the study of Volterra-type evolutionary variational inequalities is not surprising, since (6.12)–(6.13) and (5.1)–(5.2) have a common feature, (6.12)–(6.13) being a perturbation of (5.1)–(5.2) with an integral term.

The main result of this section is the following.

Theorem 6.4. *Let X be a Hilbert space and assume that* (6.2), (6.4), (6.14)–(6.17) *hold. Then there exists a unique solution u to problem* (6.12)–(6.13). *Moreover, the solution satisfies $u \in W^{1,2}(0,T;X)$.*

The proof of Theorem 6.4 will be carried out in several steps and is based on Theorem 5.12 and a fixed point argument. We assume in what follows that (6.2), (6.4), (6.14)–(6.17) hold and we introduce the set

$$\mathcal{W} = \{\eta \in W^{1,2}(0,T;X) \mid \eta(0) = 0_X\}. \tag{6.18}$$

The first step is the following.

Lemma 6.5. *For each $\eta \in \mathcal{W}$, there exists a unique $u_\eta \in W^{1,2}(0,T;X)$ such that*

$$\begin{aligned} a(u_\eta(t), v - \dot{u}_\eta(t)) + (\eta(t), v - \dot{u}_\eta(t))_X + j(v) - j(\dot{u}_\eta(t)) \\ \geq (f(t), v - \dot{u}_\eta(t))_X \quad \forall\, v \in X, \quad \text{a.e. } t \in (0,T), \end{aligned} \tag{6.19}$$

$$u_\eta(0) = u_0. \tag{6.20}$$

Lemma 6.5 is a simple consequence of Theorem 5.12. It also can be obtained from Theorem 5.1 if the function j satisfies assumption (5.4), as well as by Theorem 5.11 if (5.63) holds and there exists a family of functionals $\{j_\rho\}$ that satisfies (5.64) and (3.17).

In the next step, we consider the operator $\Lambda : \mathcal{W} \to \mathcal{W}$ defined by

$$\Lambda\eta(t) = \int_0^t A(t-s)u_\eta(s)\,ds, \quad \forall\, \eta \in \mathcal{W}, \quad t \in [0,T]. \tag{6.21}$$

Keeping in mind assumption (6.14), it is easy to check that if $\eta \in \mathcal{W}$, then $\Lambda\eta \in \mathcal{W}$ and therefore the operator Λ is well defined. A standard computation shows that

$$\Big(\frac{d}{dt}\Lambda\eta\Big)(t) = \ A(0)u_\eta(t) + \int_0^t \dot{A}(t-s)u_\eta(s)\,ds \quad \forall\, \eta \in \mathcal{W}, \quad \text{a.e. } t \in (0,T). \tag{6.22}$$

We have the following result.

Lemma 6.6. *The operator Λ has a unique fixed point $\eta^* \in \mathcal{W}$.*

Proof. Let $\eta_1, \eta_2 \in \mathcal{W}$ and, for the sake of simplicity, denote $u_1 = u_{\eta_1}$ and $u_2 = u_{\eta_2}$. Using (6.21) and (6.14), it follows that

$$\|\Lambda\eta_1(t) - \Lambda\eta_2(t)\|_X^2 \leq c \int_0^t \|u_1(s) - u_2(s)\|_X^2 \, ds \quad \forall \, t \in [0, T]. \tag{6.23}$$

Here and below in this section, c represents a positive constant that may depend on $\|A\|_{W^{1,2}(0,T;\mathcal{L}(X))}$, a and T, and whose value may change from line to line.

Moreover, from (6.22) we deduce that

$$\begin{aligned}
&\left\| \left(\frac{d}{dt}\Lambda\eta_1\right)(t) - \left(\frac{d}{dt}\Lambda\eta_2\right)(t) \right\|_X \\
&\quad \leq \|A(0)\|_{\mathcal{L}(X)} \|u_1(t) - u_2(t)\|_X + \int_0^t \|\dot{A}(t-s)\|_{\mathcal{L}(X)} \|u_1(s) - u_2(s)\|_X \, ds \\
&\quad \leq c \, \|u_1(t) - u_2(t)\|_X + c \left(\int_0^t \|u_1(s) - u_2(s)\|_X^2 \, ds \right)^{\frac{1}{2}} \quad \text{a.e. } t \in (0, T),
\end{aligned}$$

which implies that

$$\begin{aligned}
&\left\| \left(\frac{d}{dt}\Lambda\eta_1\right)(t) - \left(\frac{d}{dt}\Lambda\eta_2\right)(t) \right\|_X^2 \\
&\quad \leq c \left(\|u_1(t) - u_2(t)\|_X^2 + \int_0^t \|u_1(s) - u_2(s)\|_X^2 \, ds \right) \quad \text{a.e. } t \in (0, T).
\end{aligned} \tag{6.24}$$

On the other hand, from (6.19) we have

$$\begin{aligned}
a(u_1, v - \dot{u}_1) + (\eta_1, v - \dot{u}_1)_X + j(v) - j(\dot{u}_1) &\geq (f, v - \dot{u}_1)_X, \\
a(u_2, v - \dot{u}_2) + (\eta_2, v - \dot{u}_2)_X + j(v) - j(\dot{u}_2) &\geq (f, v - \dot{u}_2)_X,
\end{aligned}$$

for all $v \in X$, a.e. on $(0, T)$. Choose $v = \dot{u}_2$ in the first inequality, $v = \dot{u}_1$ in the second inequality, and add the results to obtain

$$\frac{1}{2} \frac{d}{dt} a(u_1 - u_2, u_1 - u_2) \leq -(\eta_1 - \eta_2, \dot{u}_1 - \dot{u}_2)_X \quad \text{a.e. on } (0, T).$$

Let $t \in [0, T]$ be fixed. Integrating the previous inequality from 0 to t and using (6.20), we find

$$\begin{aligned}
\frac{1}{2} a(u_1(t) - u_2(t), u_1(t) - u_2(t)) &\leq -(\eta_1(t) - \eta_2(t), u_1(t) - u_2(t))_X \\
&\quad + \int_0^t (\dot{\eta}_1(s) - \dot{\eta}_2(s), u_1(s) - u_2(s))_X \, ds.
\end{aligned}$$

It follows now from (6.2) that

$$\frac{m}{2}\,\|u_1(t)-u_2(t)\|_X^2 \le \|\eta_1(t)-\eta_2(t)\|_X\|u_1(t)-u_2(t)\|_X$$
$$+\int_0^t \|\dot\eta_1(s)-\dot\eta_2(s)\|_X\|u_1(s)-u_2(s)\|_X\,ds$$

and, using the inequality

$$ab \le \frac{a^2}{2\alpha} + 2\alpha b^2 \quad \forall\, a,\, b,\, \alpha > 0$$

with a convenient choice of α, we find

$$\|u_1(t)-u_2(t)\|_X^2 \le c\bigg(\|\eta_1(t)-\eta_2(t)\|_X^2 + \int_0^t \|\dot\eta_1(s)-\dot\eta_2(s)\|_X^2 ds$$
$$+\int_0^t \|u_1(s)-u_2(s)\|_X^2\,ds\bigg). \tag{6.25}$$

Now, since

$$\eta_1(t)-\eta_2(t) = \int_0^t (\dot\eta_1(s)-\dot\eta_2(s))\,ds,$$

we deduce that

$$\|\eta_1(t)-\eta_2(t)\|_X^2 \le c\int_0^t \|\dot\eta_1(s)-\dot\eta_2(s)\|_X^2\,ds.$$

Substituting this inequality in (6.25), we obtain

$$\|u_1(t)-u_2(t)\|_X^2 \le c\bigg(\int_0^t \|\dot\eta_1(s)-\dot\eta_2(s)\|_X^2\,ds + \int_0^t \|u_1(s)-u_2(s)\|_X^2\,ds\bigg).$$

Applying now Gronwall's inequality (Lemma 1.44) yields

$$\|u_1(t)-u_2(t)\|_X^2 \le c\int_0^t \|\dot\eta_1(s)-\dot\eta_2(s)\|_X^2\,ds, \tag{6.26}$$

which implies that

$$\int_0^t \|u_1(s)-u_2(s)\|_X^2\,ds \le c\int_0^t \|\dot\eta_1(s)-\dot\eta_2(s)\|_X^2\,ds. \tag{6.27}$$

We combine (6.23), (6.24), (6.26), and (6.27) to obtain

$$\|\Lambda\eta_1(t)-\Lambda\eta_2(t)\|_X^2 + \left\|\Big(\frac{d}{dt}\Lambda\eta_1\Big)(t) - \Big(\frac{d}{dt}\Lambda\eta_2\Big)(t)\right\|_X^2$$
$$\le c\int_0^t \|\dot\eta_1(s)-\dot\eta_2(s)\|_X^2\,ds.$$

Iterating the last inequality, we infer

$$\|\Lambda^m\eta_1(t) - \Lambda^m\eta_2(t)\|_X^2 + \left\| \left(\frac{d}{dt}\Lambda^m\eta_1\right)(t) - \left(\frac{d}{dt}\Lambda^m\eta_2\right)(t) \right\|_X^2$$

$$\leq c^m \int_0^t \int_0^{s_1} \cdots \int_0^{s_{m-1}} \|\dot{\eta}_1(s_m) - \dot{\eta}_2(s_m)\|_X^2 \, ds_m \cdots ds_1,$$

where Λ^m denotes the m-th power of the operator Λ. The last inequality implies

$$\|\Lambda^m\eta_1 - \Lambda^m\eta_2\|_{W^{1,2}(0,T;X)}^2 \leq \frac{c^m\,T^m}{m!} \|\eta_1 - \eta_2\|_{W^{1,2}(0,T;X)}^2.$$

Since $\lim\limits_{m\to\infty} \dfrac{c^m T^m}{m!} = 0$, the previous inequality implies that for m large enough, a power Λ^m of Λ is a contraction in $\mathcal{W}$. Lemma 6.6 is now a consequence of Theorem 1.41, since $\mathcal{W}$ is a closed subspace of the Hilbert space $W^{1,2}(0,T;X)$. □

We have now all the ingredients needed to prove the theorem.

Proof. Existence. Let $\eta^* \in \mathcal{W}$ be the fixed point of Λ and let u_{η^*} be the function defined by Lemma 6.5 for $\eta = \eta^*$. Since $\Lambda\eta^* = \eta^*$, it follows from (6.19)–(6.21) that u_{η^*} is a solution of problem (6.12)–(6.13). Moreover, the regularity $u_{\eta^*} \in W^{1,2}(0,T;X)$ is obtained from Lemma 6.5.

Uniqueness. The uniqueness is a consequence of the uniqueness of the fixed point of the operator Λ. Indeed, let u be a solution of problem (6.12)–(6.13) with regularity $u \in W^{1,2}(0,T;X)$ and consider the element $\eta \in \mathcal{W}$ defined by

$$\eta(t) = \int_0^t A(t-s)u(s)\,ds \quad \forall\, t \in [0,T]. \tag{6.28}$$

It follows that u is the solution of problem (6.19)–(6.20), and by the uniqueness of this solution (guaranteed by Lemma 6.5) we have

$$u = u_\eta. \tag{6.29}$$

We conclude by (6.21), (6.28), and (6.29) that $\Lambda\eta = \eta$; therefore, from the uniqueness of the fixed point of the operator Λ (guaranteed by Lemma 6.6), it follows that

$$\eta = \eta^*. \tag{6.30}$$

The uniqueness part of Theorem 6.4 is now a consequence of (6.29) and (6.30). □

We end this section with the following result.

Corollary 6.7. *Let X be a Hilbert space and let $K \subset X$ be a nonempty, convex, and closed subset. Assume that* (6.2), (6.14)–(6.16) *hold and, moreover,* *there exists* $\delta_0 \geq 0$ *such that* $a(u_0, v) \geq (f(0), v)_X - \delta_0$ *for all* $v \in K$. *Then there exists a unique function* $u \in W^{1,2}(0,T;X)$ *such that* $u(0) = u_0$ *and*

$$\dot{u}(t) \in K, \quad a(u(t), v - \dot{u}(t)) + \left(\int_0^t A(t-s)\, u(s)\, ds, v - \dot{u}(t) \right)_X \\ \geq (f(t), v - \dot{u}(t))_X \quad \forall\, v \in K, \;\; a.e.\; t \in (0,T).$$

Corollary 6.7 is a direct consequence of Theorem 6.4 and is obtained by choosing $j = \psi_K$, where ψ_K represents the indicator function of the set K.

6.3 Convergence Results

In this section, we study the dependence of the solution of the Volterra-type variational inequalities on the operator A and on the function j, as well.

We start with the case of elliptic variational inequalities. To this end, we assume in what follows that (6.2)–(6.5) hold and we denote by $u \in C([0,T];X)$ the solution of (6.1) obtained in Theorem 6.1. Let Θ be a given set of parameters and, for every $\theta \in \Theta$, denote by A_θ a perturbation of A that satisfies

$$A_\theta \in C([0,T]; \mathcal{L}(X)). \tag{6.31}$$

Also, consider the problem of finding $u_\theta : [0,T] \to X$ such that

$$a(u_\theta(t), v - u_\theta(t)) + \left(\int_0^t A_\theta(t-s)\, u_\theta(s)\, ds, v - u_\theta(t) \right)_X + j(v) - j(u_\theta(t)) \\ \geq (f(t), v - u_\theta(t))_X \quad \forall\, v \in X, \;\; t \in [0,T]. \tag{6.32}$$

We deduce from Theorem 6.1 that the variational inequality (6.32) has a unique solution $u_\theta \in C([0,T];X)$, for every $\theta \in \Theta$. Consider now the assumption

$$\|A_\theta - A\|_{C([0,T];\mathcal{L}(X))} \to 0. \tag{6.33}$$

We have the following convergence result.

Theorem 6.8. *Assume that* (6.2)–(6.5), (6.31), *and* (6.33) *hold. Then the solution* u_θ *of problem* (6.32) *converges to the solution* u *of problem* (6.1), *i.e.,*

$$\|u_\theta - u\|_{C([0,T];X)} \to 0. \tag{6.34}$$

Note that the convergence result presented in Theorem 6.8 is understood in the following sense, already described on pages 48 and 95: for every sequence $\{\theta_n\} \subset \Theta$ such that $\|A_{\theta_n} - A\|_{C([0,T];\mathcal{L}(X))} \to 0$ as $n \to \infty$, one has $\|u_{\theta_n} - u\|_{C([0,T];X)} \to 0$ as $n \to \infty$.

Proof. Let $\theta \in \Theta$. Everywhere below c will represent a positive constant that may depend on a, A, u, and T, but is independent of t and θ, and whose value may change from place to place. Denote

$$\eta_\theta(t) = \int_0^t A_\theta(t-s)\, u_\theta(s)\, ds \quad \forall\, t \in [0,T], \tag{6.35}$$

$$\eta(t) = \int_0^t A(t-s)\, u(s)\, ds \quad \forall\, t \in [0,T]. \tag{6.36}$$

We use (6.32) and (6.1) to obtain

$$a(u_\theta(t), v - u_\theta(t)) + (\eta_\theta(t), v - u_\theta(t))_X + j(v) - j(u_\theta(t)) \geq (f(t), v - u_\theta(t))_X,$$
$$a(u(t), v - u(t)) + (\eta(t), v - u(t))_X + j(v) - j(u(t)) \geq (f(t), v - u(t))_X,$$

for all $v \in X$ and $t \in [0,T]$, and using arguments similar to those in the proof of (6.9) yields

$$\|u_\theta(t) - u(t)\|_X \leq c\, \|\eta_\theta(t) - \eta(t)\|_X \quad \forall t \in [0,T]. \tag{6.37}$$

It follows from (6.35) and (6.36) that

$$\begin{aligned}\|\eta_\theta(t) - \eta(t)\|_X &\leq \int_0^t \|A_\theta(t-s)u_\theta(s) - A(t-s)u(s)\|_X\, ds \\ &\leq \int_0^t \|A_\theta(t-s)\|_{\mathcal{L}(X)} \|u_\theta(s) - u(s)\|_X\, ds \\ &\quad + \int_0^t \|A_\theta(t-s) - A(t-s)\|_{\mathcal{L}(X)} \|u(s)\|_X\, ds \quad \forall\, t \in [0,T].\end{aligned}$$

So,

$$\begin{aligned}\|\eta_\theta(t) - \eta(t)\|_X \leq \|A_\theta\|_{C([0,T];\mathcal{L}(X))} \int_0^t \|u_\theta(s) - u(s)\|_X\, ds \\ + c\, \|A_\theta - A\|_{C([0,T];\mathcal{L}(X))} \quad \forall\, t \in [0,T].\end{aligned} \tag{6.38}$$

As $\theta \to 0$, it follows from (6.33) that there exists $c > 0$ such that

$$\|A_\theta\|_{C([0,T];\mathcal{L}(X))} \leq c. \tag{6.39}$$

We combine (6.37), (6.38), and (6.39) to obtain

$$\|u_\theta(t) - u(t)\|_X \leq c \left(\int_0^t \|u_\theta(s) - u(s)\|_X\, ds + \|A_\theta - A\|_{C([0,T];\mathcal{L}(X))} \right)$$

for all $t \in [0,T]$, and using Gronwall's inequality (Lemma 1.44), we find that

$$\|u_\theta(t) - u(t)\|_X \leq c\, \|A_\theta - A\|_{C([0,T];\mathcal{L}(X))} \quad \forall\, t \in [0,T]. \tag{6.40}$$

The convergence result (6.34) is now a consequence of (6.33) and (6.40). $\square$

We turn now to the case of evolutionary variational inequalities. To this end, we assume in what follows that (6.2), (6.4), (6.14)–(6.17) hold and we denote by $u \in W^{1,2}(0,T;X)$ the solution of (6.12)–(6.13) obtained in Theorem 6.4. Let again Θ be a given set of parameters and, for every $\theta \in \Theta$, denote by A_θ a perturbation of A that satisfies

$$A_\theta \in W^{1,2}(0,T;\mathcal{L}(X)). \tag{6.41}$$

We consider the problem of finding $u_\theta : [0,T] \to X$ such that

$$a(u_\theta(t), v - \dot{u}_\theta(t)) + \left(\int_0^t A_\theta(t-s)\, u_\theta(s)\, ds, v - \dot{u}_\theta(t)\right)_X + j(v) - j(\dot{u}_\theta(t))$$
$$\geq (f(t), v - \dot{u}_\theta(t))_X \quad \forall\, v \in X, \text{ a.e. } t \in (0,T), \tag{6.42}$$

$$u_\theta(0) = u_0. \tag{6.43}$$

We deduce from Theorem 6.4 that problem (6.42)–(6.43) has a unique solution $u_\theta \in W^{1,2}(0,T;X)$, for every $\theta \in \Theta$. Consider now the assumption

$$\|A_\theta - A\|_{W^{1,2}(0,T;\mathcal{L}(X))} \to 0. \tag{6.44}$$

We have the following convergence result.

Theorem 6.9. *Assume that* (6.2), (6.4), (6.14)–(6.17), (6.41), *and* (6.44) *hold. Then the solution u_θ of problem* (6.42)–(6.43) *converges to the solution u of problem* (6.12)–(6.13), *i.e.,*

$$\|u_\theta - u\|_{C([0,T];X)} \to 0. \tag{6.45}$$

Note that the convergence result in Theorem 6.9 is understood, again, in the sense described on page 117.

Proof. Let $\theta \in \Theta$. Everywhere below c will represent a positive constant that may depend on a, A, u, and T, but is independent of t and θ, and whose value may change from place to place. Denote again

$$\eta_\theta(t) = \int_0^t A_\theta(t-s) u_\theta(s)\, ds \quad \forall\, t \in [0,T], \tag{6.46}$$

$$\eta(t) = \int_0^t A(t-s) u(s)\, ds \quad \forall\, t \in [0,T]. \tag{6.47}$$

Keeping in mind (6.42) and (6.12), we write

$$a(u_\theta, v - \dot{u}_\theta) + (\eta_\theta, v - \dot{u}_\theta)_X + j(v) - j(\dot{u}_\theta) \geq (f, v - \dot{u}_\theta)_X,$$
$$a(u, v - \dot{u}) + (\eta, v - \dot{u}) + j(v) - j(\dot{u}) \geq (f, v - \dot{u})_X,$$

for all $v \in X$, a.e. on $(0,T)$, and using arguments similar to those used in the proof of (6.26) yields

$$\|u_\theta(t) - u(t)\|_X^2 \le c \int_0^t \|\dot{\eta}_\theta(s) - \dot{\eta}(s)\|_X^2 ds \quad \forall\, t \in [0,T]. \tag{6.48}$$

It follows now from (6.46) and (6.47) that

$$\dot{\eta}_\theta(t) = A_\theta(0) u_\theta(t) + \int_0^t \dot{A}_\theta(t-s) u_\theta(s)\, ds,$$
$$\dot{\eta}(t) = A(0) u(t) + \int_0^t \dot{A}(t-s) u(s)\, ds,$$

a.e. $t \in (0,T)$, which imply

$$\begin{aligned}\|\dot{\eta}_\theta(t) - \dot{\eta}(t)\|_X \le \|A_\theta(0)\|_{\mathcal{L}(X)} \|u_\theta(t) - u(t)\|_X + \|A_\theta(0) - A(0)\|_{\mathcal{L}(X)} \|u(t)\|_X \\ + \int_0^t \|\dot{A}_\theta(t-s)\|_{\mathcal{L}(X)} \|u_\theta(s) - u(s)\|_X\, ds \\ + \int_0^t \|\dot{A}_\theta(t-s) - \dot{A}(t-s)\|_{\mathcal{L}(X)} \|u(s)\|_X\, ds, \end{aligned} \tag{6.49}$$

a.e. $t \in (0,T)$. As $\theta \to 0$, it follows from (6.44) that there exists $c > 0$ such that

$$\|A_\theta\|_{C([0,T];\mathcal{L}(X))} \le c, \quad \|\dot{A}_\theta\|_{L^2(0,T;\mathcal{L}(X))} \le c.$$

Using now these inequalities in (6.49) will yield after some algebra

$$\begin{aligned}&\|\dot{\eta}_\theta(t) - \dot{\eta}(t)\|_X^2 \\ &\le c\Big(\|A_\theta - A\|_{W^{1,2}(0,T;\mathcal{L}(X))}^2 + \|u_\theta(t) - u(t)\|_X^2 + \int_0^t \|u_\theta(s) - u(s)\|_X^2 ds\Big),\end{aligned}$$

a.e. $t \in (0,T)$. The previous inequality implies

$$\begin{aligned}&\int_0^t \|\dot{\eta}_\theta(s) - \dot{\eta}(s)\|_X^2\, ds \\ &\le c\left(\|A_\theta - A\|_{W^{1,2}(0,T;\mathcal{L}(X))}^2 + \int_0^t \|u_\theta(s) - u(s)\|_X^2\, ds\right) \quad \forall\, t \in [0,T].\end{aligned} \tag{6.50}$$

We combine now (6.48) and (6.50) to obtain

$$\|u_\theta(t) - u(t)\|_X^2 \le c\left(\|A_\theta - A\|_{W^{1,2}(0,T;\mathcal{L}(X))}^2 + \int_0^t \|u_\theta(s) - u(s)\|_X^2\, ds\right)$$

for all $t \in [0,T]$ and, using again Gronwall's inequality, we find

$$\|u_\theta(t) - u(t)\|_X \le c\, \|A_\theta - A\|_{W^{1,2}(0,T;\mathcal{L}(X))} \quad \forall\, t \in [0,T]. \tag{6.51}$$

The convergence (6.45) is now a consequence of (6.44) and (6.51). □

We now continue with the study of the dependence of the solution of the variational inequality (6.1) on the function j. To this end, we assume in what follows that (6.2)–(6.5) hold and, again, we denote by $u \in C([0,T];X)$ the solution of (6.1) obtained in Theorem 6.1. Also, for every $\rho > 0$, denote by j_ρ a perturbation of j that satisfies (6.4), i.e.,

$$j_\rho : X \to (-\infty, \infty] \text{ is a proper convex l.s.c. function.} \tag{6.52}$$

It follows from Theorem 6.1 that there exists a unique function $u_\rho \in C([0,T];X)$ such that

$$\begin{aligned} a(u_\rho(t), v - u_\rho(t)) + \left(\int_0^t A(t-s)u_\rho(s)\,ds, v - u_\rho(t)\right)_X + j_\rho(v) - j_\rho(u_\rho(t)) \\ \geq (f(t), v - u_\rho(t))_X \quad \forall\, v \in X,\ t \in [0,T]. \end{aligned} \tag{6.53}$$

The behavior of the solution u_ρ as $\rho \to 0$ is given by the following theorem.

Theorem 6.10. *Assume that* (6.2)–(6.5), (6.52), *and* (3.17) *hold. Then the solution* u_ρ *of problem* (6.53) *converges to the solution* u *of problem* (6.1), *i.e.,*

$$\|u_\rho - u\|_{C([0,T];X)} \to 0 \quad \text{as } \rho \to 0. \tag{6.54}$$

Proof. Let $\rho > 0$ and $t \in [0,T]$. Everywhere below c will represent a positive constant that may depend on a, A, u, and T, but it is independent on t and ρ, and whose value may change from place to place.

We choose $v = u_\rho(t)$ in (6.1) and $v = u(t)$ in (6.53) and add the resulting inequalities to obtain

$$\begin{aligned} &a(u_\rho(t) - u(t), u_\rho(t) - u(t)) \\ &\quad \leq \left(\int_0^t A(t-s)(u(s) - u_\rho(s))\,ds, u_\rho(t) - u(t)\right)_X \\ &\qquad + j(u_\rho(t)) - j(u(t)) + j_\rho(u(t)) - j_\rho(u_\rho(t)). \end{aligned} \tag{6.55}$$

Assumption (6.3) on A implies

$$\begin{aligned} &\left(\int_0^t A(t-s)\,(u(s) - u_\rho(s))\,ds, u_\rho(t) - u(t)\right)_X \\ &\leq \left(\int_0^t \|A(t-s)\|_{\mathcal{L}(X)} \|u_\rho(s) - u(s)\|_X\,ds\right) \|u_\rho(t) - u(t)\|_X \\ &\leq c\left(\int_0^t \|u_\rho(s) - u(s)\|_X\,ds\right) \|u_\rho(t) - u(t)\|_X \end{aligned}$$

and, using the elementary inequality

$$c\,ab \le \frac{c^2}{2m}\,a^2 + \frac{m}{2}\,b^2,$$

we find that

$$\begin{aligned}&\left(\int_0^t A(t-s)\,(u(s)-u_\rho(s))\,ds, u_\rho(t)-u(t)\right)_X\\&\quad\le \frac{c^2}{2m}\left(\int_0^t \|u_\rho(s)-u(s)\|_X\,ds\right)^2 + \frac{m}{2}\,\|u_\rho(t)-u(t)\|_X^2.\end{aligned}\tag{6.56}$$

On the other hand, assumption (3.17)(a) on j_ρ leads to

$$j(u_\rho(t)) - j(u(t)) + j_\rho(u(t)) - j_\rho(u_\rho(t)) \le 2\,F(\rho).\tag{6.57}$$

We combine now inequalities (6.55)–(6.57) and use the X-ellipticity of the form a, (6.2)(b), to obtain

$$\frac{m}{2}\,\|u_\rho(t)-u(t)\|_X^2 \le \frac{c^2}{2m}\left(\int_0^t \|u_\rho(s)-u(s)\|_X\,ds\right)^2 + 2\,F(\rho)$$

which yields

$$\|u_\rho(t)-u(t)\|_X^2 \le c\left(\int_0^t \|u_\rho(s)-u(s)\|_X^2 ds + F(\rho)\right).\tag{6.58}$$

The Gronwall argument in (6.58) implies that

$$\|u_\rho(t)-u(t)\|_X \le c\,\sqrt{F(\rho)}.\tag{6.59}$$

The convergence result (6.54) is now a consequence of inequality (6.59) combined with assumption (3.17)(b). □

A convergence result similar to that in Theorem 6.10 can be obtained in the study of the evolutionary variational inequality (6.12)–(6.13). To present it, we assume in what follows that (6.2), (6.4), (6.14)–(6.17) hold and we denote by $u \in W^{1,2}(0,T;X)$ the solution of (6.12)–(6.13) obtained in Theorem 6.4. Also, for every $\rho > 0$, denote by j_ρ a perturbation of j that satisfies (6.52) and (3.17). It follows from the proof of Lemma 5.10 that

$$a(u_0,v) + j_\rho(v) - (f(0),v)_X \ge -\delta_0 - F(\rho) \quad \forall\, v \in X$$

and, therefore, by using again Theorem 6.4, we obtain that there exists a unique function $u_\rho \in W^{1,2}(0,T;X)$ such that

$$\begin{aligned}&a(u_\rho(t), v-\dot{u}_\rho(t)) + \left(\int_0^t A(t-s)\,u_\rho(s)\,ds, v-\dot{u}_\rho(t)\right)_X + j_\rho(v) - j_\rho(\dot{u}_\rho(t))\\&\quad\ge (f(t), v-\dot{u}_\rho(t))_X \quad \forall\, v \in X,\ \text{a.e. } t \in (0,T),\end{aligned}\tag{6.60}$$

$$u_\rho(0) = u_0.\tag{6.61}$$

The behavior of the solution u_ρ as $\rho \to 0$ is given by following theorem.

Theorem 6.11. *Assume that* (6.2), (6.4), (6.14)–(6.17), (6.52), *and* (3.17) *hold. Then the solution* u_ρ *of problem* (6.60)–(6.61) *converges to the solution* u *of problem* (6.12)–(6.13), *i.e.,*

$$\|u_\rho - u\|_{C([0,T];X)} \to 0 \quad \text{as } \rho \to 0. \tag{6.62}$$

Proof. We use arguments similar to those used in the proofs of Theorems 6.9 and 6.10. Since the modifications are straightforward, we omit the details.

Let $\rho > 0$ and $t \in [0,T]$. We use (6.12), (6.60), and (3.17)(b) to obtain

$$a(u_\rho - u, \dot{u}_\rho - \dot{u}) \leq (\eta - \eta_\rho, \dot{u}_\rho - \dot{u}) + 2F(\rho) \quad \text{a.e. on } (0,T), \tag{6.63}$$

where

$$\eta_\rho(\tau) = \int_0^\tau A(\tau - s)u_\rho(s)\, ds, \quad \eta(\tau) = \int_0^\tau A(\tau - s)u(s)\, ds \tag{6.64}$$

for all $\tau \in [0,T]$. We now employ in (6.63) an argument similar to that used to obtain (6.26) and find the inequality

$$\|u_\rho(t) - u(t)\|_X^2 \leq c \left(\int_0^t \|\dot{\eta}_\rho(s) - \dot{\eta}(s)\|_X^2\, ds + F(\rho) \right). \tag{6.65}$$

Next, from (6.64) and arguments similar to those used to obtain (6.50), we obtain

$$\int_0^t \|\dot{\eta}_\rho(s) - \dot{\eta}(s)\|_X^2\, ds \leq c \int_0^t \|u_\rho(s) - u(s)\|_X^2\, ds. \tag{6.66}$$

We now combine inequalities (6.65) and (6.66) and apply the Gronwall lemma to the resulting inequality together with assumption (3.17)(b) to obtain the convergence result (6.62). □

Bibliographical Notes

Interest in variational inequalities originates in mechanical problems, see for instance [42], where many problems in mechanics and physics are formulated in the framework of variational inequalities. More recent references in the mathematical analysis of variational inequalities and nonlinear partial differential equations include [9, 12, 39, 46, 49, 85, 87, 113]. Details concerning the numerical analysis of variational inequalities can be found in [52, 53, 61, 67, 86]. A reference in the study of mathematical and numerical analysis of variational inequalities arising in hardening plasticity is [57].

The results presented in Section 3.1 are standard and can be found in many books and surveys. In the proof of Theorem 3.1, we followed arguments that can be found in [44, 85]. A different proof of Theorem 3.1 can be obtained following the ideas in [15]; there, the main ingredients in the proof are the properties of the proximity maps, introduced in [105], and the Banach fixed point theorem. Section 3.2 was written following results that can be found in [53, Ch. 1]. The proof of Theorem 3.7, based on the Banach fixed point argument, follows [28]; there, a more general existence and uniqueness result for elliptic quasivariational inequalities involving strongly monotone Lipschitz continuous operators is presented. Theorem 3.10 represents a simplified version of a more general result on elliptic quasivariational inequalities obtained in [107]. There, the existence of the solution was proved by using the properties of the directional derivative of the function j and topological degree arguments; moreover, an application in the study of a static problem modeling the frictional contact with normal compliance and Coulomb's law was considered.

The results presented in Sections 4.1 and 4.3 in the study of evolutionary variational inequalities with viscosity are based on more general results derived in [58, 59], see also [8]. In [58, 59], the unique solvability of variational inequalities of the form (4.29)–(4.30) and (4.31)–(4.32) was proved together with their numerical analysis, based on the study of semidiscrete and fully discrete schemes. Error estimates and convergence results were also obtained. Moreover, applications in the study of three-dimensional frictional contact problems with normal compliance and normal damped response were provided. The results in Sections 4.2 and 4.4 represent abstract versions of the results obtained in [3, 24, 122, 129] in the study of variational inequalities modeling quasistatic frictional contact problems involving Kelvin-Voigt viscoelastic materials. A survey of these results can be found in [60]. There, the existence and uniqueness of the weak solution for various viscoelastic and viscoplastic contact problems is proved, fully discrete numerical schemes are proposed and implemented, and the results of the corresponding numerical simulations are presented.

Evolutionary variational inequalities of the form (5.1)–(5.2) have been studied by many authors, under various assumptions on the function j, see, e.g., [16, 37, 57, 60]. The results in [16] are based on arguments of nonlinear equations with maximal monotone operators. The results in [57] and [60] use a time discretization method, under the assumption that j is a positively homogeneous function or a seminorm, respectively. The time discretization method in the study of evolutionary variational inequalities arising in contact mechanics was also used in [2, 4, 29, 30, 31, 32]. In writing Sections 5.1 and 5.2, we used some ideas of [37]. There, evolutionary variational inequalities of the form (5.1)–(5.2) were considered in the case when the function j is time-dependent and the existence of the solution was proved by using a regularization method involving Gâteaux differentiable functions. Note also that a regularization method similar to that used in Section 5.2 was used in [42] in the study of parabolic evolutionary variational inequalities. A convergence result similar to that presented in Theorem 5.14 was obtained in [132], in the study of evolutionary variational inequalities with viscosity involving strongly monotone Lipschitz continuous operators. Section 5.4 was written following [106]. There, Theorem 5.15 was presented in the more general case when j is a subadditive and positively homogeneous functional; also, an application to the study of a quasistatic frictional contact problem with normal compliance for linearly elastic materials was considered.

Most of the results presented in Chapter 6 of this book are based on our research and were obtained in [123, 136]. Thus, Volterra-type elliptic variational inequalities of the form (6.1) were studied in [123] where the proof of Theorem 6.1 was obtained; the numerical approximation of the problem was also considered, based on spatially semidiscrete and fully discrete schemes, and error estimates were derived; the abstract results were then used in the study of the Signorini frictionless contact problem between two viscoelastic bodies with long memory. The study of evolutionary variational inequality (6.12)–(6.13) was performed in [136]; there, Theorems 6.4 and 6.9 were proved and then used in the study of a frictional contact problem for linearly viscoelastic materials with long memory; the numerical analysis of the corresponding frictional contact problem was performed in [124, 125].

Part III
Background on Contact Mechanics

Chapter 7

Modeling of Contact Processes

In this chapter, we present a general description of mathematical modeling of the processes involved in contact between a deformable body and an obstacle or a foundation. Such kind of processes abound in industry and everyday life and, for this reason, a considerable effort has been made in their modeling, analysis, and numerical simulations. We present the physical setting, the variables that determine the state of the system, the material behavior that is reflected in a constitutive law, the input data, the equation of evolution for the state of the system, and the boundary conditions for the system variables. In particular, we provide a description of the frictional contact conditions, including versions of the Coulomb law of dry friction and its regularizations. Most of the notions and results we present here are standard and can be found in many books on mechanics and, therefore, we skip many of the details. In this chapter, all variables are assumed to have sufficient degree of smoothness consistent with developments they are involved in.

7.1 Physical Setting

A large variety of situations involving contact phenomena can be cast in the general physical setting shown in Figure 7.1 and described in what follows. A deformable body occupies, in the reference configuration, an open bounded connected set $B \subset \mathbb{R}^3$ with boundary ∂B. We denote vectors and tensors by boldface letters, such us the position vector $\boldsymbol{x} = (x_i) \in B \cup \partial B$. Here and below, the indices i, j, k, l run from 1 to 3; an index that follows a comma indicates a derivative with respect to the corresponding component of the spatial variable $\boldsymbol{x}$, and the summation convention over repeated indices is adopted. We denote by $\mathbb{S}^3$ the space of second-order symmetric tensors on $\mathbb{R}^3$ or, equivalently, the space of symmetric matrices of order 3. The canonical

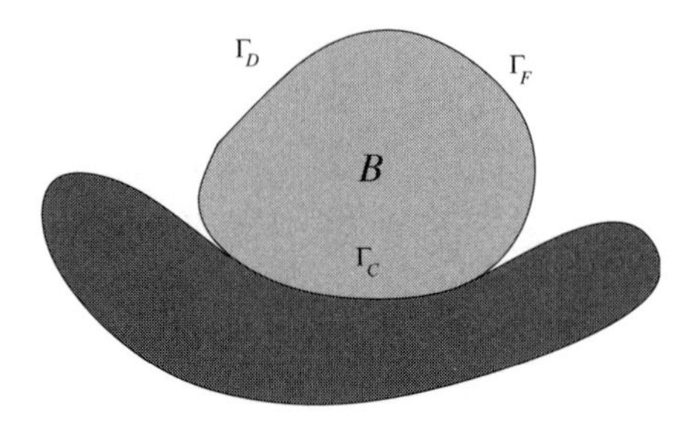

Fig. 7.1 A deformable body in contact with a foundation.

inner products and the corresponding norms on $\mathbb{R}^3$ and $\mathbb{S}^3$ are

$$\boldsymbol{u}\cdot\boldsymbol{v} = u_i v_i, \quad \|\boldsymbol{v}\| = (\boldsymbol{v}\cdot\boldsymbol{v})^{1/2} \quad \forall\, \boldsymbol{u},\boldsymbol{v}\in\mathbb{R}^3,$$
$$\boldsymbol{\sigma}\cdot\boldsymbol{\tau} = \sigma_{ij}\tau_{ij}, \quad \|\boldsymbol{\tau}\| = (\boldsymbol{\tau}\cdot\boldsymbol{\tau})^{1/2} \quad \forall\, \boldsymbol{\sigma},\boldsymbol{\tau}\in\mathbb{S}^3,$$

respectively.

The boundary ∂B is assumed to be composed of three sets $\overline{\Gamma}_D$, $\overline{\Gamma}_F$, and $\overline{\Gamma}_C$, with mutually disjoint relatively open sets Γ_D, Γ_F, and Γ_C. We assume the boundary ∂B is Lipschitz continuous and therefore the unit outward normal vector $\boldsymbol{\nu}$ exists a.e. on ∂B. The body is clamped on Γ_D. Surface tractions of density $\boldsymbol{f}_2$ act on Γ_F and volume forces of density $\boldsymbol{f}_0$ act in B. The body is in contact on Γ_C with an obstacle, the so-called foundation. The force densities $\boldsymbol{f}_0$ and $\boldsymbol{f}_2$ may depend on time. As a result, the mechanical state of the body evolves on the time interval $[0,T]$, where $T>0$.

We are interested in providing a mathematical model that describes the evolution of the body in this physical setting. To this end, we denote by $\boldsymbol{\sigma} = \boldsymbol{\sigma}(\boldsymbol{x},t) = (\sigma_{ij}(\boldsymbol{x},t))$ the *stress field* and by $\boldsymbol{u} = \boldsymbol{u}(\boldsymbol{x},t) = (u_i(\boldsymbol{x},t))$ the *displacement field.* The functions $\boldsymbol{u} : B\times[0,T]\to\mathbb{R}^3$ and $\boldsymbol{\sigma} : B\times[0,T]\to\mathbb{S}^3$ will play the role of the unknowns of the contact problem. From time to time, we suppress the explicit dependence of the quantities on the spatial variable $\boldsymbol{x}$, or both $\boldsymbol{x}$ and t; i.e., when it is convenient to do so, we write $\boldsymbol{\sigma}(t)$ and $\boldsymbol{u}(t)$, or even $\boldsymbol{\sigma}$ and $\boldsymbol{u}$. Also, everywhere below a dot above a variable will represent the derivative with respect to time and, therefore, $\dot{\boldsymbol{u}}$ denotes the velocity field.

We assume that the volume forces and surface tractions change slowly in time so that the inertia of the mechanical system is negligible. In other words, we neglect the acceleration term in the equation of motion and therefore we consider a *quasistatic process* of contact. Thus, the stress field satisfies the equation of equilibrium

$$\operatorname{Div}\boldsymbol{\sigma} + \boldsymbol{f}_0 = \boldsymbol{0} \quad \text{in } B\times(0,T) \tag{7.1}$$

where Div is the divergence operator, that is, $\operatorname{Div}\boldsymbol{\sigma} = (\sigma_{ij,j})$, and recall that $\sigma_{ij,j} = \frac{\partial\sigma_{ij}}{\partial x_j}$. Since the body is clamped on Γ_D, we impose the displacement boundary condition

$$\boldsymbol{u} = \boldsymbol{0} \quad \text{on } \Gamma_D \times (0,T). \tag{7.2}$$

The traction boundary condition is

$$\boldsymbol{\sigma\nu} = \boldsymbol{f}_2 \quad \text{on } \Gamma_F \times (0,T). \tag{7.3}$$

It states that the stress vector $\boldsymbol{\sigma\nu}$ is given on part Γ_F of the boundary during the contact process.

On Γ_C we will specify contact conditions. The contact may be frictionless or frictional, with a rigid or with a deformable foundation and, in the case of frictional contact, a number of different friction conditions may be employed. A survey of the various contact and frictional conditions used in the literature will be provided in Sections 7.3–7.5.

We now recall the strain-displacement relation

$$\boldsymbol{\varepsilon}(\boldsymbol{u}) = (\varepsilon_{ij}(\boldsymbol{u})), \quad \varepsilon_{ij}(\boldsymbol{u}) = \frac{1}{2}\,(u_{i,j} + u_{j,i}) \quad \text{in } B \times (0,T) \tag{7.4}$$

which defines the infinitesimal strain tensor. Sometimes, we omit the explicit dependence of $\boldsymbol{\varepsilon}$ on $\boldsymbol{u}$ by writing $\boldsymbol{\varepsilon}$ instead of $\boldsymbol{\varepsilon}(\boldsymbol{u})$. Note that all the problems studied in this book are formulated in the framework of small strain theory.

At this stage, the description of our model is not complete: we have more unknown functions than the number of equations. Indeed, equations (7.1) and (7.4) represent, when written out in their component forms, a total of nine equations: three from the equilibrium equation and six from strain-displacement relations (taking into account the symmetry of $\boldsymbol{\varepsilon}$). The total number of unknowns is fifteen: three components of the displacement, six components of the strain, and six components for the stress (again accounting for the symmetry of $\boldsymbol{\varepsilon}$ and $\boldsymbol{\sigma}$). Thus, it is clear that six additional equations are required if we are to have a problem that is at least in principle solvable.

Physical considerations also indicate that the description of the problem so far is incomplete. In addition to the kinematics and the balance laws that apply to all the materials, we also need description of the particular material behavior. This information, embodied in the *constitutive equation* or *constitutive law* of the material, will provide the remaining equations of the problem. A description of the constitutive laws considered in this book will be given in Section 7.2. Here we restrict ourselves to remark that in this book, we deal only with elastic and viscoelastic materials.

To conclude, the physical setting of the contact problems described above is modeled by the system of partial differential equations (7.1), (7.4), supplemented by the constitutive equation that describes the physical properties of the material; the boundary conditions associated to this system are (7.2), (7.3), and the contact condition that is formulated on $\Gamma_C \times (0,T)$; finally, if necessary, the system is completed with initial conditions for the displacement field or/and the stress field. Note that in the case of static processes for elastic materials, the data $\boldsymbol{f}_0$ and $\boldsymbol{f}_2$ do not depend on time and, since the time variable is not involved in the constitutive law and in the contact boundary conditions, relations (7.1), (7.4) hold in B, and conditions (7.2) and (7.3) hold on Γ_D and Γ_F, respectively.

7.2 Constitutive Laws

As we noted in Section 7.1, equations (7.1) and (7.4) do not constitute a complete description of the evolution of a continuous body. To obtain a complete model, valid for a given material, we must add the constitutive law of the material. Although they must satisfy some basic axioms and invariance principles, constitutive laws originate mostly from experience. A general description of several diagnostic experiments that are made with "standard" universal testing machines can be found in [36, 41, 60]. Below we just present a description of the elastic and viscoelastic constitutive laws used in this book, based on rheological considerations.

In rheology, elastic properties are described by *springs*, i.e., elements that provide a one-to-one connection between the stress σ and the strain ε. A linearly elastic spring obeys *Hooke's law*,

$$\sigma = E\,\varepsilon, \tag{7.5}$$

where the coefficient $E > 0$ is called the *Young modulus*. Viscous properties are modeled by *dashpots*, i.e., elements that provide a one-to-one connection between the stress σ and the strain rate $\dot{\varepsilon}$. A linear dashpot provides a linear relation between the stress and the strain rate, that is,

$$\sigma = \eta\,\dot{\varepsilon}. \tag{7.6}$$

Equation (7.6) is called *Newton's constitutive law*, and the coefficient $\eta > 0$ is called the *Newtonian viscosity*. Recall that here and below, a dot above a variable denotes the derivative of that variable with respect to the time. Basic rheological elements are plotted schematically in Figure 7.2.

To proceed with the construction of rheological models, it is natural to combine these elements in groups. Two possibilities arise in combining these elements: in series and in parallel. The former leads to the *Maxwell model*, see Figure 7.3, whereas the latter leads to the *Kelvin-Voigt model*, see Figure 7.4.

To obtain more general models, we combine the basic elements (springs and dashpots) as well as the Maxwell and Kelvin-Voigt elements in parallel and series. For example, taking the Maxwell model together with an elastic spring in parallel, we obtain the *standard viscoelastic model*, see Figure 7.5. By this approach, an arbitrary number of constitutive models may be constructed

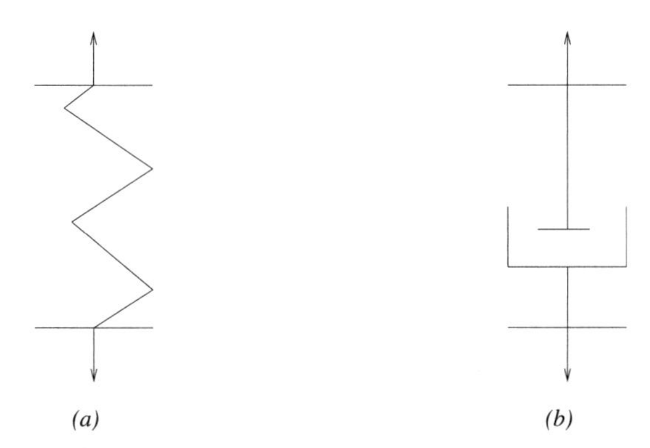

Fig. 7.2 Basic rheological elements: (a) spring; (b) dashpot.

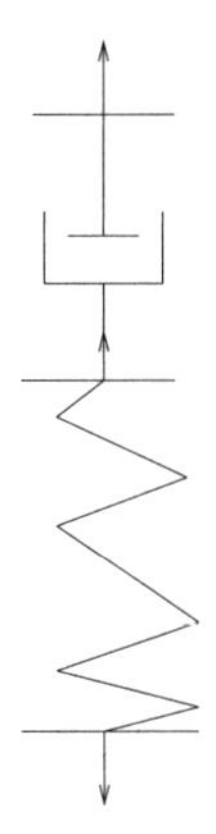

Fig. 7.3 The Maxwell model.

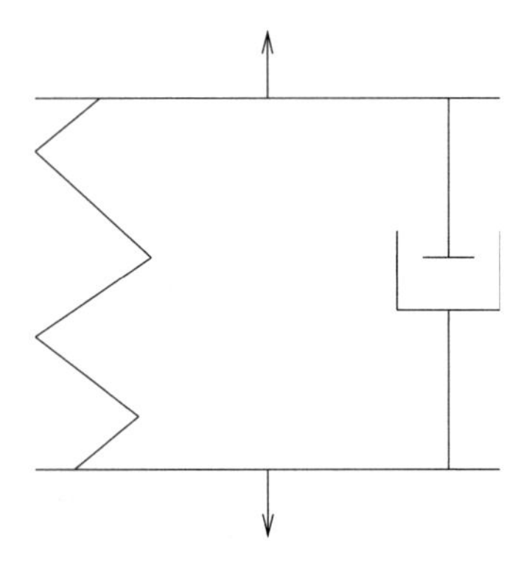

Fig. 7.4 The Kelvin-Voigt model.

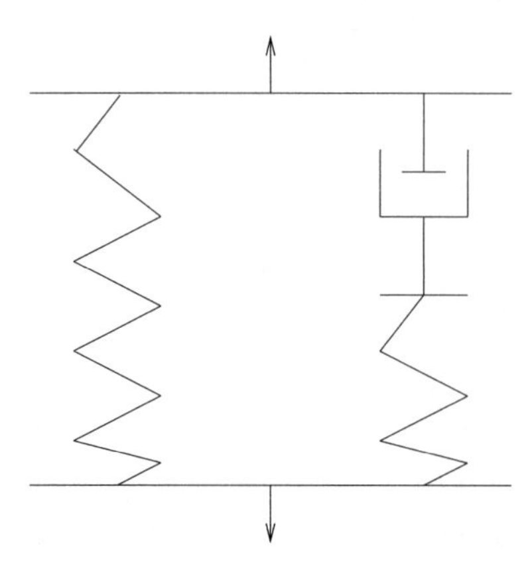

Fig. 7.5 The standard viscoelastic model.

that differ from one another by the number of elements and the way they are connected. In order to derive the differential equations corresponding with the rheological models introduced above, the following two rules are to be used:

- For a system of elements connected in series, their stresses coincide and the total strain equals the sum of strains in separate elements;
- For a system of elements connected in parallel, their strains coincide and the total stress equals the sum of stresses in separate elements.

Let us apply these rules to the Maxwell model presented in Figure 7.3. We find

$$\varepsilon = \varepsilon_e + \varepsilon_v, \tag{7.7}$$

where ε_e and ε_v are the strains in the spring and in the dashpot. Assume the rheological elements are linear; then, by using (7.5) and (7.6), we have

$$\dot{\varepsilon}_e = \frac{\dot{\sigma}}{E}, \quad \dot{\varepsilon}_v = \frac{\sigma}{\eta}. \tag{7.8}$$

It follows now from (7.7) and (7.8) that

$$\dot{\varepsilon} = \frac{\dot{\sigma}}{E} + \frac{\sigma}{\eta}, \tag{7.9}$$

which represents the constitutive law of the linear Maxwell viscoelastic material.

We now apply the above procedure to the Kelvin-Voigt model described in Figure 7.4. We have

$$\sigma = \sigma_e + \sigma_v, \tag{7.10}$$

where σ_e and σ_v are the stresses in the spring and in the dashpot, respectively. Assume the rheological elements are linear; by using (7.5) and (7.6), we have

$$\sigma_e = E\,\varepsilon, \quad \sigma_v = \eta\,\dot{\varepsilon}. \tag{7.11}$$

Then by (7.10) and (7.11) we have

$$\sigma = E\,\varepsilon + \eta\,\dot{\varepsilon}, \tag{7.12}$$

which represents the constitutive law of the linear Kelvin-Voigt viscoelastic material.

We now derive the constitutive equation for the linear standard viscoelastic solid shown in Figure 7.5. We have

$$\sigma = \sigma_e + \sigma_m, \tag{7.13}$$

where σ_e is the stress in the spring and σ_m is the stress in the Maxwell element. Since the constitutive elements are assumed to be linear, using (7.5) and (7.9) we have

$$\sigma_e = E_1 \varepsilon, \quad \dot{\varepsilon} = \frac{\dot{\sigma}_m}{E} + \frac{\sigma_m}{\eta}, \tag{7.14}$$

where E_1 denotes the Young modulus of the spring, whereas E and η represent the Young modulus and the Newtonian viscosity for the Maxwell element. We now eliminate σ_e and σ_m from (7.13) and (7.14) to obtain

$$\frac{\dot{\sigma}}{E} + \frac{\sigma}{\eta} = \left(1 + \frac{E_1}{E}\right) \dot{\varepsilon} + \frac{E_1}{\eta}\,\varepsilon, \tag{7.15}$$

which represents the constitutive equation of the linear standard viscoelastic material. Clearly, the Maxwell model (7.9) can be recovered from the standard viscoelastic constitutive law (7.15) as $E_1 = 0$.

We now write equation (7.15) in a different form. To this end, we consider a new variable $z = z(s)$ given by

$$z(s) = \sigma(s)\, e^{\frac{E}{\eta}\, s}. \tag{7.16}$$

It follows that

$$\dot{z}(s) = E \left[\frac{\dot{\sigma}(s)}{E} + \frac{\sigma(s)}{\eta}\right] e^{\frac{E}{\eta}\, s}$$

and, using (7.15), we find

$$\dot{z}(s) = \left[(E + E_1)\,\dot{\varepsilon}(s) + \frac{E\,E_1}{\eta}\,\varepsilon(s)\right] e^{\frac{E}{\eta}\, s}.$$

We now integrate the previous equation with the initial condition $z(0) = \sigma(0) = 0$ to obtain

$$z(t) = \int_0^t \left[(E + E_1)\,\dot{\varepsilon}(s) + \frac{E\,E_1}{\eta}\,\varepsilon(s) \right] e^{\frac{E}{\eta}\,s}\,ds$$

and, performing an integration by parts with the initial condition $\varepsilon(0) = 0$, after some elementary manipulations we deduce that

$$z(t) = (E + E_1)\varepsilon(t)e^{\frac{E}{\eta}\,t} - \frac{E^2}{\eta}\int_0^t \varepsilon(s)\,e^{\frac{E}{\eta}\,s}\,ds. \tag{7.17}$$

We combine now (7.16) and (7.17) to see that

$$\sigma(t) = z(t)\,e^{-\frac{E}{\eta}\,t} = (E + E_1)\varepsilon(t) - \frac{E^2}{\eta}\int_0^t \varepsilon(s)\,e^{-\frac{E}{\eta}\,(t-s)}\,ds. \tag{7.18}$$

Next, with the notation

$$E_0 = E + E_1, \quad b(s) = -\frac{E^2}{\eta}\,e^{-\frac{E}{\eta}\,s},$$

from (7.18) we deduce the following integral expression for the stress:

$$\sigma(t) = E_0\,\varepsilon(t) + \int_0^t b(t-s)\,\varepsilon(s)\,ds. \tag{7.19}$$

The constant E_0 is the *Young modulus* of the standard viscoelastic solid and the function b is the *relaxation function.* Equation (7.19) shows that the stress $\sigma(t)$ at current instant t depends on the whole history of strains up to this moment of time. Such an equality is called a *Volterra equation* and it provides a typical example of integral constitutive models. The above procedure shows that the differential models (7.9) and (7.15) are equivalent to an integral model.

To conclude, the rheological models presented above are described either by the Hooke law (7.5), or by the Kelvin-Voigt law (7.12) or, finally, by the integral law (7.19), which recovers both the standard viscoelastic and the Maxwell constitutive laws. We now extend these laws to the three-dimensional case to obtain

$$\boldsymbol{\sigma} = \mathcal{A}\,\boldsymbol{\varepsilon}(\boldsymbol{u}), \tag{7.20}$$

$$\boldsymbol{\sigma} = \mathcal{A}\,\boldsymbol{\varepsilon}(\boldsymbol{u}) + \mathcal{B}\,\boldsymbol{\varepsilon}(\dot{\boldsymbol{u}}), \tag{7.21}$$

$$\boldsymbol{\sigma}(t) = \mathcal{A}\,\boldsymbol{\varepsilon}(\boldsymbol{u}(t)) + \int_0^t \mathcal{B}(t-s)\,\boldsymbol{\varepsilon}(\boldsymbol{u}(s))\,ds, \tag{7.22}$$

where $\mathcal{A} = (a_{ijkl})$ and $\mathcal{B} = (b_{ijkl})$ are fourth-order tensors. $\mathcal{A}$ is the tensor of *elastic coefficients* and $\mathcal{B}$ is the tensor of *viscosity coefficients* when used in (7.21) and the tensor of *relaxation coefficients* when used in (7.22). Equality (7.20) represents a linear elastic constitutive law; also, as usual in the literature (see, e.g., [42]), we refer to (7.21) as a linear viscoelastic constitutive law with *short memory* and to (7.22) as a linear viscoelastic constitutive law with *long memory.*

In component form, the constitutive equations (7.20)–(7.22) read

$$\sigma_{ij} = a_{ijkl}\,\varepsilon_{kl}(\boldsymbol{u}),$$

$$\sigma_{ij} = a_{ijkl}\,\varepsilon_{kl}(\boldsymbol{u}) + b_{ijkl}\,\varepsilon_{kl}(\dot{\boldsymbol{u}}),$$

$$\sigma_{ij}(t) = a_{ijkl}\,\varepsilon_{kl}(\boldsymbol{u}(t)) + \int_0^t b_{ijkl}(t-s)\,\varepsilon_{kl}(\boldsymbol{u}(s))\,ds,$$

respectively. We allow the coefficients a_{ijkl} and b_{ijkl} to depend on the location of the point in the body.

For symmetry reasons, there are at most 21 different coefficients in each tensor. When the material is isotropic, each tensor is characterized by only two coefficients. Thus, the constitutive law of a linearly elastic isotropic material is given by

$$\boldsymbol{\sigma} = 2\,\mu\,\boldsymbol{\varepsilon}(\boldsymbol{u}) + \lambda\,tr(\boldsymbol{\varepsilon}(\boldsymbol{u}))\,\boldsymbol{I}$$

where λ and μ are the *Lamé coefficients* and satisfy $\lambda > 0$, $\mu > 0$. Here $tr(\boldsymbol{\varepsilon}(\boldsymbol{u}))$ denotes the trace of the tensor $\boldsymbol{\varepsilon}(\boldsymbol{u})$,

$$tr(\boldsymbol{\varepsilon}(\boldsymbol{u})) = \varepsilon_{ii}(\boldsymbol{u}),$$

and $\boldsymbol{I}$ denotes the identity tensor on $\mathbb{R}^3$. In components, we have

$$\sigma_{ij} = 2\,\mu\,\varepsilon_{ij}(\boldsymbol{u}) + \lambda\,\varepsilon_{kk}(\boldsymbol{u})\,\delta_{ij} \tag{7.23}$$

where δ_{ij} is the Kronecker delta. The constitutive law for a linearly viscoelastic isotropic material with short memory is given by

$$\boldsymbol{\sigma} = 2\,\mu\,\boldsymbol{\varepsilon}(\boldsymbol{u}) + \lambda\,tr(\boldsymbol{\varepsilon}(\boldsymbol{u}))\,\boldsymbol{I} + 2\,\theta\,\boldsymbol{\varepsilon}(\dot{\boldsymbol{u}}) + \zeta\,tr(\boldsymbol{\varepsilon}(\dot{\boldsymbol{u}}))\,\boldsymbol{I}$$

or, in components

$$\sigma_{ij} = 2\,\mu\,\varepsilon_{ij}(\boldsymbol{u}) + \lambda\,\varepsilon_{kk}(\boldsymbol{u})\,\delta_{ij} + 2\,\theta\,\varepsilon_{ij}(\dot{\boldsymbol{u}}) + \zeta\,\varepsilon_{kk}(\dot{\boldsymbol{u}})\,\delta_{ij}. \tag{7.24}$$

Here, again, λ and μ are the Lamé coefficients whereas θ and ζ represent the *viscosity coefficients* that satisfy $\theta > 0$, $\zeta \geq 0$. Finally, the constitutive law for a linearly viscoelastic isotropic material with long memory is

$$\begin{aligned}\boldsymbol{\sigma}(t) = 2\,\mu\,\boldsymbol{\varepsilon}(\boldsymbol{u}(t)) + \lambda\,tr(\boldsymbol{\varepsilon}(\boldsymbol{u}(t)))\,\boldsymbol{I} + 2\int_0^t \theta(t-s)\,\boldsymbol{\varepsilon}(\boldsymbol{u}(s))\,ds \\ + \int_0^t \zeta(t-s)\,tr(\boldsymbol{\varepsilon}(\boldsymbol{u}(s)))\,\boldsymbol{I}\,ds\end{aligned}$$

or, in components,

$$\sigma_{ij}(t) = 2\,\mu\,\varepsilon_{ij}(\boldsymbol{u}(t)) + \lambda\,\varepsilon_{kk}(\boldsymbol{u}(t))\,\delta_{ij} + 2\int_0^t \theta(t-s)\,\varepsilon_{ij}(\boldsymbol{u}(s))\,ds$$
$$+\int_0^t \zeta(t-s)\,\varepsilon_{kk}(\boldsymbol{u}(s))\,\delta_{ij}\,ds. \tag{7.25}$$

Here θ and ζ represent *relaxation coefficients* that are time-dependent.

In the rest of the book, we consider only isotropic materials. However, note that, since we deal with the nonhomogeneous case, we allow the coefficients λ, μ, θ, and ζ above to depend on the position $\boldsymbol{x}$.

7.3 Contact Conditions

We turn now to the description of various conditions on the contact surface Γ_C. These are divided, naturally, into the conditions in the normal direction and those in the tangential directions. To describe them we denote by u_ν and $\boldsymbol{u}_\tau$ the *normal* and *tangential* components of the displacement field $\boldsymbol{u}$ on the boundary, given by

$$u_\nu = \boldsymbol{u}\cdot\boldsymbol{\nu}, \qquad \boldsymbol{u}_\tau = \boldsymbol{u} - u_\nu\boldsymbol{\nu}. \tag{7.26}$$

Similarly, the *normal* and *tangential* components of the velocity field $\dot{\boldsymbol{u}}$ on the boundary are defined by

$$\dot{u}_\nu = \dot{\boldsymbol{u}}\cdot\boldsymbol{\nu}, \qquad \dot{\boldsymbol{u}}_\tau = \dot{\boldsymbol{u}} - \dot{u}_\nu\boldsymbol{\nu}. \tag{7.27}$$

Below we refer to the tangential components $\boldsymbol{u}_\tau$ and $\dot{\boldsymbol{u}}_\tau$ as the *slip* and the *slip rate*, respectively. We also denote by σ_ν and $\boldsymbol{\sigma}_\tau$ the *normal* and *tangential* components of the stress field $\boldsymbol{\sigma}$ on the boundary, that is,

$$\sigma_\nu = (\boldsymbol{\sigma}\boldsymbol{\nu})\cdot\boldsymbol{\nu}, \qquad \boldsymbol{\sigma}_\tau = \boldsymbol{\sigma}\boldsymbol{\nu} - \sigma_\nu\boldsymbol{\nu}. \tag{7.28}$$

The component $\boldsymbol{\sigma}_\tau$ represents the *friction force* on the contact surface Γ_C.

We start with the presentation of the conditions in the normal direction, called also *contact conditions* or *contact laws*, and we present the conditions in the tangential directions in the next section.

First, we consider the so-called *bilateral contact condition*, which describes the situation when the contact between the body and the foundation is maintained at all times. It can be found in many machines and in many parts and components of mechanical equipment. In this condition, the normal displacement vanishes on the contact surface, that is,

$$u_\nu = 0 \quad \text{on } \Gamma_C \times (0,T). \tag{7.29}$$

From a mathematical point of view, this condition is very convenient since in general it leads to a linear subspace for the set of admissible displacements fields. For this reason it was used in a number of papers, for details see [60, 130] and the references therein.

Next, we consider the case when the normal stress is prescribed on the contact surface, i.e.,

$$-\sigma_\nu = F \quad \text{on } \Gamma_C \times (0,T), \tag{7.30}$$

where F is a given positive function. Such type of contact conditions arise in the study of some mechanisms and was considered by a number of authors (see, e.g., [42, 113]). It also arises in geophysics in the study of earthquake models, see for instance [20, 74, 75, 76] and the references therein.

The so-called *normal compliance contact condition* describes a reactive foundation. It assigns a reactive normal pressure that depends on the interpenetration of the asperities on the body's surface and those on the foundation. A general expression for the normal compliance condition is

$$-\sigma_\nu = p_\nu(u_\nu) \quad \text{on } \Gamma_C \times (0,T) \tag{7.31}$$

where $p_\nu(\cdot)$ is a non-negative prescribed function that vanishes for negative argument. Indeed, when $u_\nu < 0$, there is no contact, and the normal pressure vanishes. An example of the normal compliance function p_ν is

$$p_\nu(r) = c_\nu r_+, \tag{7.32}$$

or, more generally,

$$p_\nu(r) = c_\nu (r_+)^m.$$

Here the constant $c_\nu > 0$ is the surface *stiffness coefficient*, $m > 0$ is the normal exponent, and $r_+ = \max\{0, r\}$ is the positive part of r. The contact condition (7.31) was first introduced in [99, 111] and since then used in many publications, see, e.g., the references in [130]. The term *normal compliance* was first used in [88, 89].

The contact with a perfectly rigid foundation is modeled in engineering literature by the *Signorini contact condition* and can be stated as follows:

$$u_\nu \le 0, \quad \sigma_\nu \le 0, \quad \sigma_\nu u_\nu = 0 \quad \text{on } \Gamma_C \times (0,T). \tag{7.33}$$

Inequality $u_\nu \le 0$ shows that there is no penetration; the rest of the conditions in (7.33) show that if there is no contact between the body and the foundation (i.e., $u_\nu < 0$), then the foundation does not produce a reaction toward the body, since $\sigma_\nu = 0$; if there is contact (i.e., $u_\nu = 0$), then the foundation exerts a normal compression force $\sigma_\nu \le 0$ on the body. Note that the Signorini condition (7.33) is obtained, formally, from the normal compliance condition (7.31), (7.32) in the limit when the surface stiffness coefficient becomes infinite, i.e., $c_\nu \to \infty$. The Signorini contact condition was first introduced in [131] and then used in many papers, see, e.g., [130] for further details and references.

7.4 Coulomb's Law of Dry Friction

We turn now to the conditions in the tangential directions, called also *frictional conditions* or *friction laws.* The simplest one is the so-called *frictionless* condition in which the friction force vanishes during the process, i.e.,

$$\boldsymbol{\sigma}_\tau = \mathbf{0} \quad \text{on } \Gamma_C \times (0, T). \tag{7.34}$$

This is an idealization of the process, since even completely lubricated surfaces generate shear resistance to tangential motion. However, (7.34) is a sufficiently good approximation of the reality in some situations.

In the case when the friction force $\boldsymbol{\sigma}_\tau$ does not vanish on the contact surface, the contact is *frictional.* Frictional contact is usually modeled with the *Coulomb law of dry friction* or its variants. According to this law, the magnitude of the tangential traction $\boldsymbol{\sigma}_\tau$ is bounded by a function, the so-called *friction bound,* which is the maximal frictional resistance that the surface can generate; also, once slip starts, the frictional resistance opposes the direction of the motion and its magnitude reaches the friction bound. Thus,

$$\|\boldsymbol{\sigma}_\tau\| \leq g, \quad \boldsymbol{\sigma}_\tau = -\, g\, \frac{\dot{\boldsymbol{u}}_\tau}{\|\dot{\boldsymbol{u}}_\tau\|} \quad \text{if } \dot{\boldsymbol{u}}_\tau \neq \mathbf{0} \quad \text{on } \Gamma_C \times (0, T), \tag{7.35}$$

where $\dot{\boldsymbol{u}}_\tau$ is the tangential velocity or slip rate and g represents the friction bound. On a nonhomogeneous surface, g depends explicitly on the position $\boldsymbol{x}$ on the surface; it also depends on the process variables, and we describe this dependence below in this section.

Note that the Coulomb law (7.35) is characterized by the existence of stick-slip zones on the contact boundary at each time moment $t \in [0, T]$. Indeed, it follows from (7.35) that if in a point $\boldsymbol{x} \in \Gamma_C$ the inequality $\|\boldsymbol{\sigma}_\tau(\boldsymbol{x}, t)\| < g(\boldsymbol{x})$ holds, then $\dot{\boldsymbol{u}}_\tau(\boldsymbol{x}, t) = \mathbf{0}$ and the material point $\boldsymbol{x}$ is in the so-called *stick zone*; if $\|\boldsymbol{\sigma}_\tau(\boldsymbol{x}, t)\| = g(\boldsymbol{x})$, then the point $\boldsymbol{x}$ is in the so-called *slip zone.* We conclude that Coulomb's friction law (7.35) models the phenomenon that slip may occur only when the magnitude of the friction force reaches a critical value, the friction bound g.

In certain applications, especially where the bodies are light or the friction is very large, the function g in (7.35) does not depend on the process variables and behaves like a function that depends only on the position $\boldsymbol{x}$ on the contact surface. Considering

$$g = g(\boldsymbol{x}), \tag{7.36}$$

in (7.35) leads to the *Tresca friction law*, and it simplifies considerably the analysis of the corresponding contact problem, see for instance [60, 130].

Often, especially in engineering literature, the friction bound g is chosen as

$$g = g(\sigma_\nu) = \widetilde{\mu}\,|\sigma_\nu| \tag{7.37}$$

where $\widetilde{\mu} \geq 0$ is the *coefficient of friction*. The choice (7.37) in (7.35) leads to the *classical version of Coulomb's law*, which was intensively studied in the literature, see for instance the references in [130].

When the wear of the contacting surface is taken into account, a *modified version of Coulomb's law* is more appropriate. This law has been derived in [139, 140, 141] from thermodynamic considerations and is given by choosing

$$g = \widetilde{\mu}\,|\sigma_\nu|\,(1 - \delta|\sigma_\nu|)_+ \tag{7.38}$$

in (7.35), where δ is a very small positive parameter related to the wear constant of the surface and, again, $\widetilde{\mu}$ is the coefficient of friction.

The choice (7.30) in (7.37) leads to the friction bound

$$g = \widetilde{\mu}\,F \tag{7.39}$$

and the choice (7.30) in (7.38) leads to the friction bound

$$g = \widetilde{\mu}\,F(1 - \delta F)_+. \tag{7.40}$$

It follows from (7.39) and (7.40) that, in the case when $\widetilde{\mu}$ and δ are given, the friction bound g is a given function defined on the contact surface Γ_C, and therefore we recover the Tresca friction law.

We observe that the *friction coefficient* $\widetilde{\mu}$ is not an intrinsic thermodynamic property of a material, a body or its surface, since it depends on the contact process and the operating conditions. The issue is considerably complicated by the following facts. Engineering surfaces are not mathematically smooth surfaces but contain asperities and various irregularities. Moreover, very often they contain some or all of the following: moisture, lubrication oils, various debris, wear particles, oxide layers, chemicals and materials that are different from those of the parent body. Therefore, it is not surprising that the friction coefficient is found to depend on the surface characteristics, on the surface geometry and structure, on the relative velocity between the contacting surfaces, on the surface temperature, on the wear or rearrangement of the surface, and, therefore, on its history. A very thorough description of these issues can be found in [119] (see also the survey [142]).

Until very recently, mathematical models for frictional contact used a constant friction coefficient, mainly for mathematical reasons. This is rapidly changing, and the dependence of $\widetilde{\mu}$ on the process parameters has been incorporated into the models in recent publications. Thus, the choice

$$\widetilde{\mu} = \widetilde{\mu}\,(\|\boldsymbol{u}_\tau\|) \tag{7.41}$$

was considered in many geophysical publications to model the motion of tectonic plates (see, e.g., [20, 119, 137] and the references therein); also, the choice

$$\widetilde{\mu} = \widetilde{\mu}\,(\|\dot{\boldsymbol{u}}_\tau\|) \tag{7.42}$$

was considered by many authors, see for instance the references in [130]. Assumptions (7.41) and (7.42) allow the coefficient of friction $\widetilde{\mu}$ to depend on the slip $\boldsymbol{u}_\tau$ or on the slip rate $\dot{\boldsymbol{u}}_\tau$, respectively.

A version of the dependence (7.41) and (7.42) may be obtained by assuming that the coefficient of friction depends on the total slip or on the total slip rate of the surface, that is,

$$\widetilde{\mu} = \widetilde{\mu}\,(S\boldsymbol{u}(t)) \tag{7.43}$$

or

$$\widetilde{\mu} = \widetilde{\mu}\,(S\dot{\boldsymbol{u}}(t)), \tag{7.44}$$

respectively. Here

$$S\boldsymbol{w}(t) = \int_0^t \|\boldsymbol{w}_\tau(s)\|\,ds, \tag{7.45}$$

and the functions $S\boldsymbol{u}(\boldsymbol{x},t)$ and $S\dot{\boldsymbol{u}}(\boldsymbol{x},t)$ represent the *total slip* and the *total slip rate* at the point $\boldsymbol{x}$ on Γ_C over the time period $[0,t]$, respectively.

Assume again that the choice (7.30) is made in (7.37) or (7.38) and therefore, as seen above, the friction bounds (7.39) or (7.40) are obtained. Then, taking into account the dependence (7.41) or (7.42), we obtain that the friction bound g satisfies

$$g = g(\|\boldsymbol{u}_\tau\|) \tag{7.46}$$

or

$$g = g(\|\dot{\boldsymbol{u}}_\tau\|). \tag{7.47}$$

The previous equalities allow the friction bound g to depend on the slip $\boldsymbol{u}_\tau$ or on the slip rate $\dot{\boldsymbol{u}}_\tau$. For this reason, the friction law (7.35) associated to (7.46) will be called a *slip-dependent friction law* and the same friction law (7.35) associated to (7.47) will be called a *slip rate–dependent friction law.*

A similar argument based on the dependence (7.43) or (7.44) shows that there is a need to extend the Tresca friction law by considering the case

$$g = g(S\boldsymbol{u}(t)) \tag{7.48}$$

or

$$g = g(S\dot{\boldsymbol{u}}(t)). \tag{7.49}$$

The friction law (7.35) associated to (7.48) will be called a *total slip-dependent friction law* and the same friction law (7.35) associated to (7.49) will be called a *total slip rate–dependent friction law.*

Various different models of frictional contact may be obtained by combining the normal compliance contact condition (7.31) or the Signorini contact condition (7.33) with the Coulomb law (7.35), in which the choice of the friction bound is given by (7.37) or (7.38). For instance, combining (7.31) with (7.37) leads to the friction bound $g = \widetilde{\mu} p_\nu(u_\nu)$ and combining (7.31) with (7.38) leads to the friction bound $g = \widetilde{\mu} p_\nu(u_\nu)(1 - \delta p_\nu(u_\nu))_+$. We do not present in detail such models since they are not relevant to the study of antiplane frictional contact problems. However, we note that combining the Signorini contact condition with Coulomb's law of dry friction leads to important mathematical difficulties, see [130] for comments and details.

7.5 Regularized Friction Laws

Variational formulation of frictional contact problems with Coulomb's law leads to variational inequalities involving a nondifferentiable functional. To avoid this difficulty, several regularizations of Coulomb's law (7.35) are used in the literature, mainly for numerical reasons. Some examples are given by

$$\boldsymbol{\sigma}_\tau = -g \, \frac{\dot{\boldsymbol{u}}_\tau}{\sqrt{\|\dot{\boldsymbol{u}}_\tau\|^2 + \rho^2}} \qquad \text{on } \Gamma_C \times (0, T) \tag{7.50}$$

or

$$\boldsymbol{\sigma}_\tau = \begin{cases} -g \, \|\dot{\boldsymbol{u}}_\tau\|^{\rho - 1} \, \dot{\boldsymbol{u}}_\tau & \text{if } \ \dot{\boldsymbol{u}}_\tau \neq \boldsymbol{0} \\ \boldsymbol{0} & \text{if } \ \dot{\boldsymbol{u}}_\tau = \boldsymbol{0} \end{cases} \qquad \text{on } \Gamma_C \times (0, T). \tag{7.51}$$

We assume $\rho > 0$ in (7.50) and $0 < \rho \leq 1$ in (7.51). Note that the friction laws (7.50) and (7.51) describe situation when slip appears even for small tangential shears, which is the case when the surfaces are lubricated by a thin layer of non-Newtonian fluid. Relation (7.51) is called in the literature the *power-law friction*; indeed, in this case the tangential shear is proportional to a power of the tangential velocity and, in the particular case $\rho = 1$, (7.51) implies that the tangential shear is proportional to the tangential velocity. Also, note that the Coulomb law (7.35) is obtained, formally, from the friction laws (7.50) and (7.51) in the limit as $\rho \to 0$.

An instructive comparison of the friction laws (7.35), (7.50), and (7.51) can be made in the case of two-dimensional bodies; in this case, at each point of the contact surface there is only one tangential direction. Denote by $\boldsymbol{\tau}$ its unit vector; we have $\boldsymbol{\sigma}_\tau = \sigma_\tau \boldsymbol{\tau}$ and $\dot{\boldsymbol{u}}_\tau = \dot{u}_\tau \boldsymbol{\tau}$ where σ_τ and u_τ are real

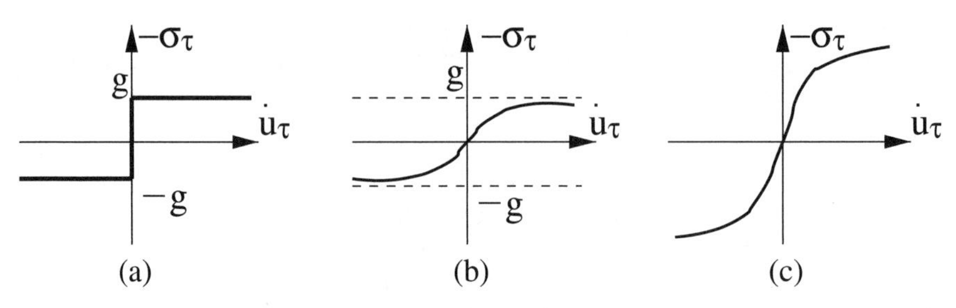

Fig. 7.6 (a) The Tresca friction law (7.52); (b) the regularized friction law (7.53); (c) the power-law friction (7.54).

valued functions. Assume that a friction bound $g \geq 0$ is given; then, in this special case, (7.35) leads to the multivalued relation

$$-\sigma_\tau = \begin{cases} -g & \text{if } \dot{u}_\tau < 0 \\ [-g,\, g] & \text{if } \dot{u}_\tau = 0 \\ g & \text{if } \dot{u}_\tau > 0, \end{cases} \tag{7.52}$$

and (7.50) and (7.51) are given by

$$-\sigma_\tau = g\,\frac{\dot{u}_\tau}{\sqrt{\dot{u}_\tau^2 + \rho^2}}, \tag{7.53}$$

$$-\sigma_\tau = \begin{cases} g\,|\dot{u}_\tau|^{\rho-1}\,\dot{u}_\tau & \text{if } \dot{u}_\tau \neq 0 \\ 0 & \text{if } \dot{u}_\tau = 0, \end{cases} \tag{7.54}$$

respectively. Relations (7.52), (7.53), and (7.54) are depicted in Figure 7.6.

Now, it is easy to see that (7.52) holds if and only if

$$-\sigma_\tau(v_\tau - \dot{u}_\tau) \leq g|v_\tau| - g|\dot{u}_\tau| \quad \forall\, v_\tau \in \mathbb{R},$$

which shows that (7.52) can be written in the subdifferential form

$$-\sigma_\tau \in \partial\varphi(\dot{u}_\tau), \tag{7.55}$$

where $\varphi : \mathbb{R} \to \mathbb{R}$ is the function

$$\varphi(s) = g|s|. \tag{7.56}$$

Note that this function is convex and continuous but it is not differentiable at $s = 0$. Also, (7.53) and (7.54) can be written in the form

$$-\sigma_\tau = \varphi'(\dot{u}_\tau) \tag{7.57}$$

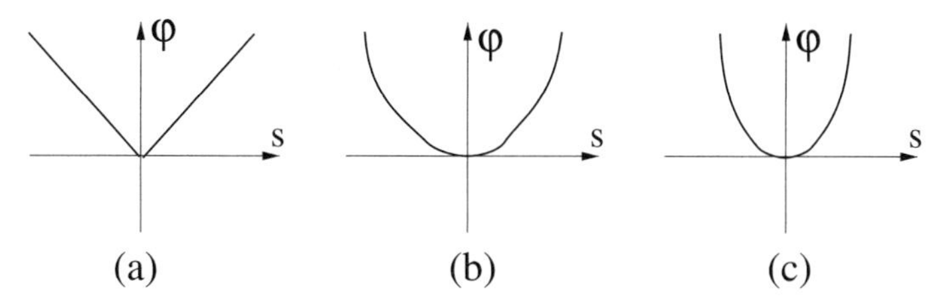

Fig. 7.7 (a) The potential $\varphi(s) = g|s|$; (b) the potential $\varphi(s) = g\sqrt{s^2+\rho^2} - \rho$; (c) the potential $\varphi(s) = \frac{g}{\rho+1}|s|^{\rho+1}$.

where φ' represents the derivative of the function $\varphi : \mathbb{R} \to \mathbb{R}$ given by

$$\varphi(s) = g\big(\sqrt{s^2+\rho^2} - \rho\big), \quad \text{in the case of (7.53).} \tag{7.58}$$

$$\varphi(s) = \frac{g}{\rho+1}\,|s|^{\rho+1}, \quad \text{in the case of (7.54).} \tag{7.59}$$

Clearly, in the last two cases above, the function φ is convex and differentiable; thus, it results from Proposition 1.39 that φ is subdifferentiable and $\partial\varphi(s) = \{\varphi'(s)\}$ for every $s \in \mathbb{R}$; therefore (7.57) can be written in the form (7.55), too. We conclude that the friction laws (7.52)–(7.54) have a common feature, as the three of them can be cast in the form (7.55) with the corresponding choice of φ. The graphs of the potential functions (7.56), (7.58), and (7.59) are plotted in Figure 7.7.

Based on the arguments and comments above, in the next chapters of this book we consider frictional contact problems involving Coulomb's law of dry friction (7.35) associated to the choice of g given by (7.36), (7.46)–(7.49). On occasions, when g satisfies (7.36), we regularize (7.35) with the friction laws (7.50) or (7.51). Note that all these laws model evolutionary frictional processes. However, sometimes we shall consider *static versions* of these laws, which are obtained by replacing the tangential velocity with the tangential displacement. Thus, the static version of the friction law (7.35) is given by

$$\|\boldsymbol{\sigma}_\tau\| \le g, \quad \boldsymbol{\sigma}_\tau = -\,g\,\frac{\boldsymbol{u}_\tau}{\|\boldsymbol{u}_\tau\|} \quad \text{if } \boldsymbol{u}_\tau \neq \boldsymbol{0} \quad \text{on } \Gamma_C. \tag{7.60}$$

The static version of the regularized friction law (7.50) is given by

$$\boldsymbol{\sigma}_\tau = -g\,\frac{\boldsymbol{u}_\tau}{\sqrt{\|\boldsymbol{u}_\tau\|^2+\rho^2}} \quad \text{on } \Gamma_C, \tag{7.61}$$

and, finally, the static version of the power-law friction (7.51) is given by

$$\boldsymbol{\sigma}_\tau = \begin{cases} -g\,\|\boldsymbol{u}_\tau\|^{\rho-1}\,\boldsymbol{u}_\tau & \text{if } \boldsymbol{u}_\tau \neq \boldsymbol{0} \\ \boldsymbol{0} & \text{if } \boldsymbol{u}_\tau = \boldsymbol{0} \end{cases} \quad \text{on } \Gamma_C. \tag{7.62}$$

The static friction laws are suitable for proportional loadings and can be considered a first approximation of the evolutionary friction laws. Considering the static versions (7.60), (7.61), and (7.62) of the friction laws (7.35), (7.50), and (7.51), respectively, simplifies the mathematical analysis of the corresponding mechanical problems.

Chapter 8
Antiplane Shear

In this chapter, we consider the mathematical modeling of a special type of process, the antiplane shear. One can induce states of antiplane shear in a solid by loading it in a special way. We rarely actually load solids so as to cause them to deform in antiplane shear. However, we will find that the governing equations and boundary conditions for antiplane shear problems are beautifully simple, and the solution will have many of the features of the more general case and may help us to solve the more complex problem, too. For this reason, in the recent years considerable attention has been paid to the analysis of antiplane shear deformation within the context of elasticity theory. We start the description of the basic assumptions, then we specialize the equilibrium and constitutive equations in the context of antiplane shear and introduce a Sobolev-type space used in the study of such kind of problems. Further, we consider an elastic antiplane displacement-traction boundary value problem, derive its variational formulation, and prove the existence of the unique weak solution by using the Lax-Milgram theorem. Finally, we present a complete description of the frictional contact conditions used in this book, in the context of antiplane process; to this end, we consider both the case of unstressed and pre-stressed reference configurations.

8.1 Basic Assumptions and Equations

We will discuss antiplane shear in the context of a three-dimensional boundary value problem. To this end, we refer in what follows to the physical setting described in Section 7.1. We assume that B is a cylinder with generators parallel to the x_3-axes having a cross section that is a regular region Ω in the x_1, x_2-plane, $Ox_1x_2x_3$ being a Cartesian coordinate system. The cylinder is assumed to be sufficiently long so that end effects in the axial direction are negligible. Thus, $B = \Omega \times (-\infty, +\infty)$. Let $\partial\Omega = \Gamma$; we assume that Γ is composed of three sets $\overline{\Gamma}_1$, $\overline{\Gamma}_2$, and $\overline{\Gamma}_3$, with the mutually disjoint relatively open sets Γ_1, Γ_2, and Γ_3, such that the one-dimensional measure of Γ_1, denoted $meas\,\Gamma_1$, is strictly positive. We choose $\Gamma_D = \Gamma_1 \times (-\infty, +\infty)$, $\Gamma_F = \Gamma_2 \times (-\infty, +\infty)$, and $\Gamma_C = \Gamma_3 \times (-\infty, +\infty)$, and, according to the description presented in Section 7.1, we assume that the cylinder is clamped on $\Gamma_1 \times (-\infty, +\infty)$ and is in contact with a rigid foundation on $\Gamma_3 \times (-\infty, +\infty)$. Moreover, the cylinder is subjected to time-dependent volume forces of density $\boldsymbol{f}_0$ in $\Omega \times (-\infty, \infty)$ and to time-dependent surface tractions of density $\boldsymbol{f}_2$ on $\Gamma_2 \times (-\infty, +\infty)$. This physical setting is depicted in Figures 8.1 and 8.2.

Let $T > 0$ and let $[0, T]$ denote the time interval of interest. We assume that

$$\boldsymbol{f}_0 = (0, 0, f_0) \quad \text{with} \quad f_0 = f_0(x_1, x_2, t) : \Omega \times [0, T] \to \mathbb{R}, \tag{8.1}$$

$$\boldsymbol{f}_2 = (0, 0, f_2) \quad \text{with} \quad f_2 = f_2(x_1, x_2, t) : \Gamma_2 \times [0, T] \to \mathbb{R}. \tag{8.2}$$

The body forces (8.1) and the surface tractions (8.2) give rise to a deformation of the elastic cylinder whose displacement $\boldsymbol{u}$ is of the form

$$\boldsymbol{u} = (0, 0, u) \quad \text{with} \quad u = u(x_1, x_2, t) : \Omega \times [0, T] \to \mathbb{R}. \tag{8.3}$$

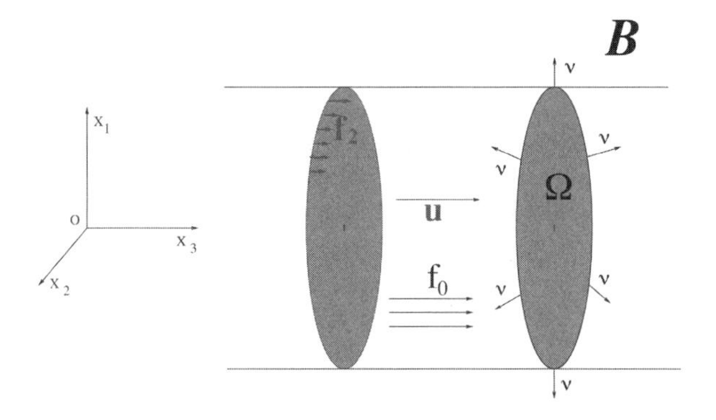

Fig. 8.1 Antiplane shear: three-dimensional view of the cylinder.

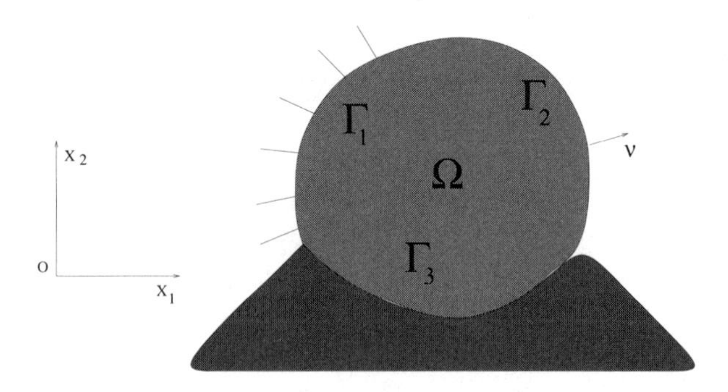

Fig. 8.2 Antiplane shear: cross section of the cylinder in contact with a foundation.

Such a deformation is called an *antiplane shear*. This displacement field may be regarded as a natural complement to that of *plane strain* where $\boldsymbol{u} = (u_1, u_2, 0)$ and $u_1 = u_1(x_1, x_2, t)$, $u_2 = u_2(x_1, x_2, t)$, see for instance [96, p. 134] for details.

From (8.3) and (7.4) we see that, in the case of the antiplane process, the infinitesimal strain tensor becomes

$$\boldsymbol{\varepsilon}(\boldsymbol{u}) = \begin{pmatrix} 0 & 0 & \frac{1}{2}\, u_{,1} \\ 0 & 0 & \frac{1}{2}\, u_{,2} \\ \frac{1}{2}\, u_{,1} & \frac{1}{2}\, u_{,2} & 0 \end{pmatrix} \tag{8.4}$$

and note that $tr(\boldsymbol{\varepsilon}(\boldsymbol{u})) = \varepsilon_{kk}(\boldsymbol{u}) = 0$.

Let $\boldsymbol{\sigma} = (\sigma_{ij})$ denote the stress field and recall that in the quasistatic processes it satisfies the equilibrium equation (7.1). Since $\mathrm{Div}\,\boldsymbol{\sigma} = (\sigma_{ij,j})$, it follows from (8.1) that

$$\left.\begin{aligned} \sigma_{11,1} + \sigma_{12,2} + \sigma_{13,3} &= 0 \\ \sigma_{21,1} + \sigma_{22,2} + \sigma_{23,3} &= 0 \\ \sigma_{31,1} + \sigma_{32,2} + \sigma_{33,3} + f_0 &= 0 \end{aligned}\right\} \quad \text{in } B \times (0, T). \tag{8.5}$$

Assume now that the cylinder is elastic and its behavior is modeled with the linear isotropic constitutive law (7.23), in which $\mu = \mu(x_1, x_2)$. We combine (7.23) and (8.4) to see that in this case $\boldsymbol{\sigma} = 2\,\mu\boldsymbol{\varepsilon}$, that is,

$$\boldsymbol{\sigma} = \begin{pmatrix} 0 & 0 & \mu u_{,1} \\ 0 & 0 & \mu u_{,2} \\ \mu u_{,1} & \mu u_{,2} & 0 \end{pmatrix}. \tag{8.6}$$

Since the functions u and f_0 depend only on the variables x_1, x_2 and t and $\mu = \mu(x_1, x_2)$, it follows from (8.6) that the first two equations in (8.5) are satisfied identically; moreover, the last equation in (8.5) becomes

$$\mathrm{div}\,(\mu\, \nabla\, u) + f_0 = 0 \quad \text{in } \Omega \times (0, T). \tag{8.7}$$

Here and everywhere in the rest of the book, we use the notation div and ∇ for the divergence and gradient operators with respect to the variables x_1 and x_2, i.e.,

$$\mathrm{div}\,\boldsymbol{\tau} = \tau_{1,1} + \tau_{2,2}, \tag{8.8}$$
$$\nabla v = (v_{,1}, v_{,2}) \tag{8.9}$$

if $\boldsymbol{\tau} = (\tau_1(x_1, x_2, t), \tau_2(x_1, x_2, t))$ or $\boldsymbol{\tau} = (\tau_1(x_1, x_2), \tau_2(x_1, x_2))$ and $v = v(x_1, x_2, t)$ or $v = v(x_1, x_2)$. We conclude that, in the case of antiplane shear of elastic materials, the equation of equilibrium reduces to the scalar equation (8.7). Note that this equation describes the nonhomogeneous case since, here and below we assume that $\mu = \mu(x_1, x_2)$. Antiplane shear deformations for a nonhomogeneous anisotropic linearly elastic cylinder were considered in [69, 71]. In the homogeneous case, equation (8.7) reduces to the time-dependent Laplace equation $\mu \, \Delta u + f_0 = 0$ in $\Omega \times (0,T)$.

Assume now that the behavior of the cylinder's material is modeled with a viscoelastic constitutive law with short memory, (7.24), in which $\mu = \mu(x_1, x_2)$ and $\theta = \theta(x_1, x_2)$. Then, using similar arguments as above, it follows that the stress tensor is given by

$$\boldsymbol{\sigma} = \begin{pmatrix} 0 & 0 & \theta \dot{u}_{,1} + \mu u_{,1} \\ 0 & 0 & \theta \dot{u}_{,2} + \mu u_{,2} \\ \theta \dot{u}_{,1} + \mu u_{,1} & \theta \dot{u}_{,2} + \mu u_{,2} & 0 \end{pmatrix} \tag{8.10}$$

and the equilibrium equation becomes

$$\operatorname{div}(\theta \, \nabla \dot{u} + \mu \, \nabla u) + f_0 = 0 \quad \text{in } \Omega \times (0,T). \tag{8.11}$$

Finally, consider the case of viscoelastic materials with long memory, (7.25), in which $\mu = \mu(x_1, x_2)$ and $\theta = \theta(x_1, x_2, t)$; then, it follows that

$$\boldsymbol{\sigma} = \begin{pmatrix} 0 & 0 & \sigma_{13} \\ 0 & 0 & \sigma_{23} \\ \sigma_{13} & \sigma_{23} & 0 \end{pmatrix}, \tag{8.12}$$

where the nonvanishing components σ_{ij} at the moment t are given by

$$\sigma_{13}(t) = \sigma_{31}(t) = \mu u_{,1}(t) + \int_0^t \theta(t-s) \, u_{,1}(s) \, ds, \tag{8.13}$$

$$\sigma_{23}(t) = \sigma_{32}(t) = \mu u_{,2}(t) + \int_0^t \theta(t-s) \, u_{,2}(s) \, ds. \tag{8.14}$$

Moreover, the equation of equilibrium reduces to the scalar equation

$$\operatorname{div}\Big(\mu \nabla u(t) + \int_0^t \theta(t-s) \nabla u(s) \, ds \Big) + f_0(t) = 0 \quad \text{in } \Omega, \ \forall \, t \in [0,T]. \tag{8.15}$$

We now discuss the displacement-traction boundary conditions. Recall that, since the cylinder is clamped on $\Gamma_1 \times (-\infty, +\infty) \times (0,T)$, the displacement field vanishes there. Thus, (8.3) implies that

$$u = 0 \quad \text{on } \Gamma_1 \times (0,T). \tag{8.16}$$

Let $\boldsymbol{\nu}$ denote the unit normal on $\Gamma \times (-\infty, +\infty)$,

$$\boldsymbol{\nu} = (\nu_1, \nu_2, 0) \quad \text{with} \quad \nu_i = \nu_i(x_1, x_2) : \Gamma \to \mathbb{R}, \ i = 1, 2. \tag{8.17}$$

Denote by $\partial_\nu u$ and $\partial_\nu \dot{u}$ the *normal derivative* of the displacement and velocity, respectively, given by

$$\partial_\nu u = \nabla u \cdot \boldsymbol{\nu} = u_{,1}\, \nu_1 + u_{,2}\, \nu_2, \tag{8.18}$$

$$\partial_\nu \dot{u} = \nabla \dot{u} \cdot \boldsymbol{\nu} = \dot{u}_{,1}\, \nu_1 + \dot{u}_{,2}\, \nu_2. \tag{8.19}$$

We use this notation to see that the Cauchy stress vector $\boldsymbol{\sigma\nu} = (\sigma_{ij}\nu_j)$ has the following form:

$$\boldsymbol{\sigma\nu} = (0, 0, \mu\, \partial_\nu u) \tag{8.20}$$

in the elastic case (8.6),

$$\boldsymbol{\sigma\nu} = (0, 0, \theta\, \partial_\nu\, \dot{u} + \mu\, \partial_\nu u) \tag{8.21}$$

in the viscoelastic case with short memory (8.10), and

$$\boldsymbol{\sigma\nu}(t) = \left(0, 0, \mu\, \partial_\nu u(t) + \int_0^t \theta(t-s)\, \partial_\nu u(s)\, ds\right) \tag{8.22}$$

in the viscoelastic case with long-term memory (8.12)–(8.14).

We use now (8.20)–(8.22) and (8.2) to see that the traction boundary condition (7.3) is given by:

$$\mu\, \partial_\nu u = f_2 \quad \text{on } \Gamma_2 \times (0, T) \tag{8.23}$$

in the case of elastic materials,

$$\theta\, \partial_\nu \dot{u} + \mu\, \partial_\nu u = f_2 \quad \text{on } \Gamma_2 \times (0, T) \tag{8.24}$$

in the case of viscoelastic materials with short memory, and, finally,

$$\mu\, \partial_\nu u(t) + \int_0^t \theta(t-s)\, \partial_\nu u(s)\, ds = f_2(t) \quad \text{on } \Gamma_2, \ \forall\, t \in [0, T] \tag{8.25}$$

in the case of viscoelastic materials with long memory.

Assume now that the body forces and surface traction do not depend on time, and the cylinder is elastic. Assume also that the displacement field u does not depend on time. Then the process is *static* and the equilibrium equation (8.7), the displacement boundary condition (8.16), and the traction boundary condition (8.23) are valid in Ω, on Γ_1, and on Γ_2, respectively.

The frictional contact conditions on Γ_3 will be discussed in Section 8.4; there, we specialize the contact conditions and the friction laws presented in Sections 7.3–7.5 in the context of antiplane problems.

8.2 A Function Space for Antiplane Problems

It follows from the previous section that the study of antiplane problems leads to second-order boundary value problems in the regular domain $\Omega \subset \mathbb{R}^2$. For this reason, in order to impose the essential boundary condition (8.16), we introduce the functional space V given by

$$V = \{ v \in H^1(\Omega) \, : \, v = 0 \text{ on } \Gamma_1 \}. \tag{8.26}$$

Here, equality $v = 0$ on Γ_1 is understood in the sense of traces, i.e., almost everywhere on Γ_1, as it was described in Section 2.3. Since the trace operator is linear and continuous, we see that V is a closed subspace of the Sobolev space $H^1(\Omega)$; therefore, combining Corollary 2.8 and Theorem 1.25, it follows that V is a separable Hilbert space with the inner product $(\cdot,\cdot)_{H^1(\Omega)}$ and the associated norm $\|\cdot\|_{H^1(\Omega)}$.

In the study of antiplane problems, it is more convenient to consider a different inner product on V, defined by

$$(u, v)_V = (\nabla u, \nabla v)_{L^2(\Omega)^2}, \tag{8.27}$$

where $(\cdot,\cdot)_{L^2(\Omega)^2}$ denotes the inner product on the space $L^2(\Omega)^2$, (2.3). Also, denote by $\|\cdot\|_V$ the associated norm, that is,

$$\|v\|_V = \|\nabla v\|_{L^2(\Omega)^2} \quad \forall\, v \in V. \tag{8.28}$$

We have the following result.

Theorem 8.1. *The space V is a separable Hilbert space with the inner product* (8.27).

Proof. We apply Corollary 2.17 to see that there exists a constant $c > 0$, which depends on Ω and Γ_1, such that

$$\|v\|_{H^1(\Omega)} \le c \left(\|\nabla v\|_{L^2(\Omega)^2} + \int_{\Gamma_1} |v| \, da \right) \quad \forall\, v \in H^1(\Omega).$$

Therefore, by the definition of V we obtain that

$$\|v\|_{H^1(\Omega)} \le c \, \|\nabla v\|_{L^2(\Omega)^2} \quad \forall\, v \in V. \tag{8.29}$$

Inequality (8.29) is called the *Friedrichs-Poincaré* inequality. As a consequence of this inequality, it follows that (8.27) is an inner product on the space V and, moreover,

$$\|v\|_{H^1(\Omega)} \le c \, \|v\|_V \quad \forall\, v \in V. \tag{8.30}$$

On the other hand, note that (8.28) and (2.10) imply that

$$\|v\|_V \le \|v\|_{H^1(\Omega)} \quad \forall\, v \in V. \tag{8.31}$$

We combine now (8.30) and (8.31) to see that $\|\cdot\|_V$ and $\|\cdot\|_{H^1(\Omega)}$ are equivalent norms on the space V, which concludes the proof. □

Now, we combine (2.11) and (8.30) to see that there exists $c > 0$, which depends on Ω and Γ_1, such that

$$\|v\|_{L^2(\Gamma)} \leq c\,\|v\|_V \quad \forall\, v \in V; \tag{8.32}$$

this implies that

$$\|v\|_{L^2(\Gamma_3)} \leq c_0\,\|v\|_V \quad \forall\, v \in V, \tag{8.33}$$

where now $c_0 > 0$ depends on Ω, Γ_1, and Γ_3. Inequality (8.33) will be used in several occasions in Part IV of the book in order to estimate various quantities defined on the contact surface Γ_3.

We end this section with a Green-type formula that will be repeatedly used in the rest of the book in order to derive variational formulations of antiplane contact problems. To this end we recall (8.8), (8.9) and we note that, with the notation made there, the equalities below hold:

$$\boldsymbol{\tau} \cdot \boldsymbol{\nu} = \tau_1\nu_1 + \tau_2\nu_2, \qquad \boldsymbol{\tau} \cdot \nabla\, v = \tau_1 v_{,1} + \tau_2 v_{,2}. \tag{8.34}$$

We have the following result.

Theorem 8.2. *Let* $\boldsymbol{\tau} = (\boldsymbol{\tau}_1, \boldsymbol{\tau}_2) \in H^1(\Omega)^2$. *Then*

$$\int_\Omega \boldsymbol{\tau} \cdot \nabla v\, dx + \int_\Omega \operatorname{div} \boldsymbol{\tau}\, v\, dx = \int_\Gamma \boldsymbol{\tau} \cdot \boldsymbol{\nu}\, v\, da \quad \forall\, v \in H^1(\Omega). \tag{8.35}$$

Proof. Let $v \in H^1(\Omega)$. First, we note that

$$\int_\Omega w\, v_{,i}\, dx + \int_\Omega w_{,i}\, v\, dx = \int_\Gamma w\, v\, \nu_i\, da \quad \forall\, w \in H^1(\Omega),\ i = 1, 2. \tag{8.36}$$

This equality can be proved first for $C^\infty(\overline{\Omega})$ functions by an integration by parts and then extended to $H^1(\Omega)$ by a density argument. We use (8.36) for $i = 1$ and $w = \tau_1$, then for $i = 2$ and $w = \tau_2$, add the resulting equalities and use notation (8.8), (8.9), and (8.34) to obtain (8.35). □

Note that (8.35) implies that

$$\int_\Omega \boldsymbol{\tau} \cdot \nabla v\, dx + \int_\Omega \operatorname{div} \boldsymbol{\tau}\, v\, dx = \int_{\Gamma_2} \boldsymbol{\tau} \cdot \boldsymbol{\nu}\, v\, da + \int_{\Gamma_3} \boldsymbol{\tau} \cdot \boldsymbol{\nu}\, v\, da \quad \forall\, v \in V, \tag{8.37}$$

since the boundary integral is split over Γ_1, Γ_2, and Γ_3 and every element $v \in V$ vanishes on Γ_1.

8.3 An Elastic Antiplane Boundary Value Problem

In this section, we provide a first existence and uniqueness result in the study of antiplane problems. We assume that the material is elastic and $\Gamma_3=\emptyset$, that is, we consider a purely displacement-traction problem. We also assume that the process is static and therefore the problem can be formulated as follows.

Problem 8.3. *Find a displacement field $u : \Omega \to \mathbb{R}$ such that*

$$\operatorname{div}(\mu \nabla u) + f_0 = 0 \quad \text{in } \Omega, \tag{8.38}$$

$$u = 0 \quad \text{on } \Gamma_1, \tag{8.39}$$

$$\mu\, \partial_\nu u = f_2 \quad \text{on } \Gamma_2. \tag{8.40}$$

Note that (8.38), (8.39), and (8.40) represent the equilibrium equation (8.7), the displacement boundary condition (8.16), and the traction boundary condition (8.23), respectively, in their static version. Once the displacement field that solves Problem 8.3 is known, then the associated stress tensor $\boldsymbol{\sigma}$ can be calculated by using (8.6). Also, note that in the homogeneous case (i.e., in the case when μ does not depend on x_1, x_2), Problem 8.3 reduces to a simple Dirichlet-Neumann problem for the Laplace equation.

Problem 8.3 is called the *classical formulation* of the elastic antiplane displacement-traction problem. In its formulation, it was assumed that all the functions used are as smooth as is needed for the various mathematical operations to be justified. A function $u : \Omega \to \mathbb{R}$ is a *classical solution* to Problem 8.3 if u is a regular function (say $u \in C^2(\Omega)\cap C^1(\overline{\Omega})$) and it satisfies (8.38) at each point $\boldsymbol{x} \in \Omega$, (8.39) at each point $\boldsymbol{x} \in \Gamma_1$, and (8.40) at each point $\boldsymbol{x} \in \Gamma_2$.

To allow for the equations and conditions to be satisfied in a "weaker sense" we need to reformulate the problem in the so-called *variational formulation*, which we proceed to do. To this end we use below the space V (page 152) together with the inner product $(\cdot,\cdot)_V$ and norm $\|\cdot\|_V$ and we start with the following result.

Lemma 8.4. *(1) Let u be a solution to Problem 8.3 such that $u \in V$ and $\mu \nabla u \in H^1(\Omega)^2$. Then $f_0 \in L^2(\Omega)$, $f_2 \in L^2(\Gamma_2)$ and u satisfies the variational equation*

$$\int_\Omega \mu \nabla u \cdot \nabla v \, dx = \int_\Omega f_0\, v\, dx + \int_{\Gamma_2} f_2\, v\, da \quad \forall\, v \in V. \tag{8.41}$$

(2) Conversely, let $f_0 \in L^2(\Omega)$, $f_2 \in L^2(\Gamma_2)$ and consider a solution u of the variational equation (8.41) such that $u \in V$ and $\mu \nabla u \in H^1(\Omega)^2$. Then u verifies the partial differential equation (8.38) in $L^2(\Omega)$ and the boundary conditions (8.39) and (8.40) in $L^2(\Gamma_1)$ and $L^2(\Gamma_2)$, respectively.

Proof. (1) Suppose that u is a solution to Problem 8.3 such that $u \in V$ and $\mu \nabla u \in H^1(\Omega)^2$. It follows from (8.38) that $f_0 = -\mathrm{div}\,(\mu \nabla u)$ and, since $\mathrm{div}\,(\mu \nabla u) \in L^2(\Omega)$, we deduce that $f_0 \in L^2(\Omega)$. Also, it follows from (8.40) and (8.18) that $f_2 = \mu\, \partial_\nu u = (\mu \nabla u) \cdot \boldsymbol{\nu} \in L^2(\Gamma_2)$. Next, to derive (8.41) we use Green's formula (8.35) to obtain

$$\int_\Omega \mu \nabla u \cdot \nabla v\, dx + \int_\Omega \mathrm{div}\,(\mu \nabla u)\, v\, dx = \int_\Gamma \mu \nabla u \cdot \boldsymbol{\nu}\, v\, da \quad \forall\, v \in V;$$

then, using (8.38) and (8.18), we find

$$\int_\Omega \mu \nabla u \cdot \nabla v\, dx = \int_\Omega f_0\, v\, dx + \int_\Gamma \mu\, \partial_\nu u\, v\, da \quad \forall\, v \in V.$$

We split the boundary integral over Γ_1 and Γ_2 and, since v vanishes on Γ_1 and $\mu\, \partial_\nu\, u = f_2$ on Γ_2, we deduce that u satisfies (8.41).

(2) Conversely, assume that u satisfies (8.41) and $u \in V$, $\mu \nabla u \in H^1(\Omega)^2$. Clearly, u satisfies condition (8.39) in the sense of traces, that is in $L^2(\Gamma_1)$, and, since $\mu \nabla u \in H^1(\Omega)^2$ and $u = 0$ on Γ_1, Green's formula (8.37) yields:

$$\int_\Omega \mu \nabla u \cdot \nabla v\, dx + \int_\Omega \mathrm{div}\,(\mu \nabla u)\, v\, dx = \int_{\Gamma_2} \mu\, \partial_\nu u\, v\, da \quad \forall\, v \in V. \tag{8.42}$$

We subtract (8.42) and (8.41) to find

$$\int_\Omega (\mathrm{div}\,(\mu \nabla u) + f_0)\, v\, dx = \int_{\Gamma_2} (\mu\, \partial_\nu u - f_2)\, v\, da \quad \forall\, v \in V. \tag{8.43}$$

We choose now $v = \varphi \in C_0^\infty(\Omega)$ in (8.43) to see that

$$\int_\Omega (\mathrm{div}\,(\mu \nabla u) + f_0)\, \varphi\, dx = 0 \quad \forall\, \varphi \in C_0^\infty(\Omega)$$

and, since the space $C_0^\infty(\Omega)$ is dense in $L^2(\Omega)$, we deduce by the previous equality that (8.38) holds in $L^2(\Omega)$. We now combine (8.38) and (8.43) to see that

$$\int_{\Gamma_2} (\mu\, \partial_\nu\, u - f_2)\, v\, da = 0 \quad \forall\, v \in V$$

and, since the set of traces of functions in V is a dense subspace in $L^2(\Gamma_2)$, it results that (8.40) holds in $L^2(\Gamma_2)$. □

Following Lemma 8.4, we focus our interest in the study of the variational equation (8.41). To this end, we assume that the body forces and tractions have the regularity

$$f_0 \in L^2(\Omega), \quad f_2 \in L^2(\Gamma_2) \tag{8.44}$$

and the Lamé coefficient satisfies

$$\begin{aligned}&\mu \in L^\infty(\Omega) \text{ and there exists } \mu^* > 0 \text{ such that}\\&\qquad \mu(\boldsymbol{x}) \geq \mu^* \text{ a.e. } \boldsymbol{x} \in \Omega.\end{aligned} \tag{8.45}$$

We denote by $a : V \times V \to \mathbb{R}$ the bilinear form

$$a(u, v) = \int_\Omega \mu \, \nabla u \cdot \nabla v \, dx \tag{8.46}$$

and we note that assumption (8.45) implies that the integral in (8.46) is well defined. We also note that by assumption (8.44), the integrals on the right-hand side of (8.41) are well defined and, moreover, the functional

$$v \mapsto \int_\Omega f_0 \, v \, dx + \int_{\Gamma_2} f_2 \, v \, da$$

is a linear continuous functional on V. We use now Riesz's representation theorem (page 11) to define an element $f \in V$ by the equality

$$(f, v)_V = \int_\Omega f_0 \, v \, dx + \int_{\Gamma_2} f_2 \, v \, da. \tag{8.47}$$

Note that (8.47) and (8.32) imply that

$$\|f\|_V \leq c \left(\|f_0\|_{L^2(\Omega)} + \|f_2\|_{L^2(\Gamma_2)}\right) \tag{8.48}$$

where c depends on Ω and Γ_1. Next, equalities (8.46), (8.47) combined with Lemma 8.4 lead us to the following formulation of Problem 8.3.

Problem 8.5. *Find a displacement field $u \in V$ such that*

$$a(u, v) = (f, v)_V \quad \forall \, v \in V. \tag{8.49}$$

It follows from the discussion in Lemma 8.4 that every smooth solution of Problem 8.3 is a solution to Problem 8.5 and conversely, if Problem 8.5 has a sufficiently smooth solution, it is also a solution of the original problem in the usual sense. However, now we have a formulation, the so-called *variational formulation*, that may have solutions that do not have the necessary regularity or smoothness, and we still call them solutions, but now we write that such solutions are *weak solutions* of the original problem.

The statement of the existence and uniqueness of the solution for Problem 8.5 is as follows.

Theorem 8.6. *Assume that* (8.44) *and* (8.45) *hold. Then Problem* 8.5 *has a unique solution $u \in V$. Moreover, there exists $c > 0$, which depends only on Ω, Γ_1, and μ^* such that*

$$\|u\|_V \leq c \left(\|f_0\|_{L^2(\Omega)} + \|f_2\|_{L^2(\Gamma_2)}\right), \tag{8.50}$$

and the solution minimizes on V the energy functional J defined by

$$J(v) = \frac{1}{2} \int_{\Omega} \mu \, \|\nabla v\|^2 \, dx - \int_{\Omega} f_0 \, v \, dx - \int_{\Gamma_2} f_2 \, v \, dx.$$

Proof. The bilinear form a defined by (8.46) is symmetric and, moreover, by (8.45) it follows that

$$|a(u,v)| \leq \|\mu\|_{L^\infty(\Omega)} \, \|u\|_V \, \|v\|_V \quad \forall \, u, \, v \in V, \tag{8.51}$$

i.e., a is continuous. It also follows (8.45) that

$$a(v,v) \geq \mu^* \, \|v\|_V^2 \quad \forall \, v \in V, \tag{8.52}$$

i.e., a is V-elliptic. The existence and uniqueness part in Theorem 8.6 is now a consequence of the Lax-Milgram theorem (Corollary 3.5, page 47). Also, we choose $v = u$ in (8.49) and use (8.52) to obtain

$$\|u\|_V \leq \frac{1}{\mu^*} \, \|f\|_V. \tag{8.53}$$

We use (8.53) and (8.48) to obtain (8.50) with a positive constant c, which depends only on Ω, Γ_1, and μ^*. The last part of the theorem follows from (8.46) and (8.47), combined with the arguments presented on page 47. □

We conclude by Theorem 8.6 that the mechanical problem 8.3 has a unique *weak solution* that depends continuously on the data.

The arguments in this section will be used in the study of the antiplane frictional contact problems presented in Part IV of the manuscript. The pattern is very similar for each problem: first, the classical formulation of the model is described, which amounts to choosing the constitutive law and the contact conditions. Then, a variational formulation of the problem is derived by performing, formally, integration by parts and using the equations and the conditions. Next, a statement of the existence and uniqueness results, under appropriate assumptions on the problem data, is provided.

8.4 Frictional Contact Conditions

In this section, we specialize the frictional contact conditions presented in Sections 7.3–7.5 to the case of antiplane shear. We first recall that for a displacement field $\boldsymbol{u}$, we denote by u_ν and $\boldsymbol{u}_\tau$ its normal and tangential components given by

$$u_\nu = \boldsymbol{u} \cdot \boldsymbol{\nu}, \quad \boldsymbol{u}_\tau = \boldsymbol{u} - u_\nu \, \boldsymbol{\nu}.$$

Therefore, by using (8.3) and (8.17), we deduce that in the antiplane context we have

$$u_\nu = 0 \tag{8.54}$$

and

$$\boldsymbol{u}_\tau = (0, 0, u). \tag{8.55}$$

Equality $u_\nu = 0$ shows that the contact is bilateral, i.e., there is no separation between the body and the foundation during the process.

Next, recall that the normal and tangential components of the stress field are given by

$$\sigma_\nu = (\boldsymbol{\sigma}\boldsymbol{\nu}) \cdot \boldsymbol{\nu}, \quad \boldsymbol{\sigma}_\tau = \boldsymbol{\sigma}\boldsymbol{\nu} - \sigma_\nu \boldsymbol{\nu}$$

and, therefore, for a stress field of the form (8.12) we have

$$\sigma_\nu = 0, \quad \boldsymbol{\sigma}_\tau = \boldsymbol{\sigma}\boldsymbol{\nu}. \tag{8.56}$$

The first equality in (8.56) shows that the normal stress vanishes on the contact boundary during the process; therefore, the Cauchy vector $\boldsymbol{\sigma}\boldsymbol{\nu}$ reduces to its tangential component $\boldsymbol{\sigma}_\tau$, as stated in the second equality in (8.56).

We now use (8.56) and (8.20)–(8.22) to see that:

$$\boldsymbol{\sigma}_\tau = (0, 0, \mu\, \partial_\nu u) \tag{8.57}$$

in the case of the elastic constitutive law (8.6),

$$\boldsymbol{\sigma}_\tau = (0, 0, \theta\, \partial_\nu\, \dot{u} + \mu\, \partial_\nu u) \tag{8.58}$$

in the case of the viscoelastic constitutive law with short memory (8.10),

$$\boldsymbol{\sigma}_\tau(t) = (0, 0, \mu\, \partial_\nu u(t) + \int_0^t \theta(t-s)\, \partial_\nu u(s)\, ds) \tag{8.59}$$

in the case of the viscoelastic constitutive law with long memory (8.12)–(8.14).

With these preliminaries, we can now describe the main frictional conditions used in Part IV of this book. First, we use (8.57)–(8.59) and (8.55) to see that the Coulomb law (7.35) reduces to the following scalar relations:

$$|\mu\, \partial_\nu\, u| \le g, \quad \mu\, \partial_\nu u = -g\, \frac{\dot{u}}{|\dot{u}|} \quad \text{if } \dot{u} \ne 0 \tag{8.60}$$

in the case of elastic materials,

$$|\theta\, \partial_\nu \dot{u} + \mu\, \partial_\nu u| \le g, \quad \theta\, \partial_\nu \dot{u} + \mu\, \partial_\nu u = -g\, \frac{\dot{u}}{|\dot{u}|} \quad \text{if } \dot{u} \ne 0 \tag{8.61}$$

in the case of viscoelastic materials with short memory,

$$\left.\begin{aligned} &\left|\mu\,\partial_\nu u(t)+\int_0^t \theta(t-s)\,\partial_\nu u(s)\,ds\right|\le g,\\ &\mu\,\partial_\nu u(t)+\int_0^t \theta(t-s)\,\partial_\nu u(s)\,ds=-g\,\frac{\dot u(t)}{|\dot u(t)|}\quad \text{if } \dot u(t)\neq 0 \end{aligned}\right\} \tag{8.62}$$

in the case of viscoelastic materials with long-term memory. Equalities (8.60) and (8.61) are satisfied on $\Gamma_3\times(0,T)$ while equality (8.62) is satisfied on Γ_3, for every $t\in[0,T]$.

Recall that the function g in (8.60)–(8.62) is called the *friction bound* and may depend on the process variables. The case when

$$g=g(\boldsymbol{x}) \tag{8.63}$$

corresponds with the nonhomogeneous Tresca friction law (7.36) where, here and below in this section, $\boldsymbol{x}=(x_1,x_2)$ denotes a generic point of Γ_3. Also, by using (8.55), (7.46), and (7.47) we see that the slip-dependent friction law is characterized by the friction bound

$$g=g(\boldsymbol{x},|u(\boldsymbol{x},\,t)|), \tag{8.64}$$

and the slip rate–dependent friction law (7.47) is characterized by the friction bound

$$g=g(\boldsymbol{x},|\dot u(\boldsymbol{x},\,t)|). \tag{8.65}$$

Using again (8.55), (7.45), (7.48), and (7.49), it follows that the total slip-dependent friction law is characterized by the friction bound

$$g=g\Big(\boldsymbol{x},\int_0^t |u(\boldsymbol{x},\,s)|\,ds\Big), \tag{8.66}$$

and that the total slip rate–dependent friction law is characterized by the friction bound

$$g=g\Big(\boldsymbol{x},\int_0^t |\dot u(\boldsymbol{x},\,s)|\,ds\Big). \tag{8.67}$$

Arguments that justify the use of the friction bounds (8.64)–(8.67) in the study of antiplane contact problems will be presented in Section 8.5 below.

In Part IV of the manuscript, we shall use the friction laws (8.60)–(8.62), associated with the choice (8.63)–(8.67) of the friction bound. When we deal with the nonhomogeneous Tresca friction law, we assume that the friction bound (8.63) satisfies

$$g\in L^2(\Gamma_3)\ \text{ and }\ g(\boldsymbol{x})\ge 0\quad \text{a.e. } \boldsymbol{x}\in\Gamma_3. \tag{8.68}$$

Also, when we use the friction bounds (8.64)–(8.67), we assume that

$$\left.\begin{array}{l}(a)\ g:\Gamma_3\times\mathbb{R}\to\mathbb{R}_+.\\ (b)\ \text{There exists } L_g>0 \text{ such that}\\ \qquad |g(\boldsymbol{x},r_1)-g(\boldsymbol{x},r_2)|\le L_g\,|r_1-r_2|\\ \qquad\quad \forall\, r_1,\,r_2\in\mathbb{R},\ \ \text{a.e. } \boldsymbol{x}\in\Gamma_3.\\ (c)\ \text{The mapping } \boldsymbol{x}\mapsto g(\boldsymbol{x},\,r)\\ \qquad \text{is Lebesgue measurable on } \Gamma_3,\ \forall\, r\in\mathbb{R}.\\ (d)\ \text{The mapping } \boldsymbol{x}\mapsto g(\boldsymbol{x},0) \text{ belongs to } L^2(\Gamma_3).\end{array}\right\}\qquad (8.69)$$

Note that condition (8.69)(b) requires that $g(\boldsymbol{x},\cdot):\mathbb{R}\to\mathbb{R}_+$ is a Lipschitz continuous function, uniformly with respect to $\boldsymbol{x}\in\Gamma_3$; the rest of the conditions in (8.69) are introduced for mathematical reasons, since they guarantee that the mapping $\boldsymbol{x}\mapsto g(\boldsymbol{x},\zeta(\boldsymbol{x}))$ belongs to $L^2(\Gamma_3)$ whenever $\zeta\in L^2(\Gamma_3)$.

Next, we use again equalities (8.57)–(8.59) and (8.55) to see that, in the antiplane shear context, the regularized friction law (7.50) reduces to the following scalar relations:

$$\mu\,\partial_\nu u=-g\,\frac{\dot{u}}{\sqrt{\dot{u}^2+\rho^2}}\qquad (8.70)$$

in the case of elastic materials,

$$\theta\,\partial_\nu\dot{u}+\mu\,\partial_\nu\,u=-g\,\frac{\dot{u}}{\sqrt{\dot{u}^2+\rho^2}}\qquad (8.71)$$

in the case of viscoelastic materials with short memory,

$$\mu\,\partial_\nu u(t)+\int_0^t\theta(t-s)\,\partial_\nu u(s)\,ds=-g\,\frac{\dot{u}(t)}{\sqrt{\dot{u}(t)^2+\rho^2}}\qquad (8.72)$$

in the case of viscoelastic materials with long memory. Here $\rho>0$ is a regularization parameter.

Also, the power-law friction (7.51) reduces to the following scalar relations:

$$\mu\,\partial_\nu u=\begin{cases}-g\,|\dot{u}|^{\rho-1}\,\dot{u} & \text{if } \dot{u}\neq 0\\ 0 & \text{if } \dot{u}=0\end{cases}\qquad (8.73)$$

in the case of elastic materials,

$$\theta\,\partial_\nu\dot{u}+\mu\,\partial_\nu u=\begin{cases}-g\,|\dot{u}|^{\rho-1}\,\dot{u} & \text{if } \dot{u}\neq 0\\ 0 & \text{if } \dot{u}=0\end{cases}\qquad (8.74)$$

in the case of viscoelastic materials with short memory,

$$\mu\,\partial_\nu u(t)+\int_0^t\theta(t-s)\,\partial_\nu u(s)\,ds=\begin{cases}-g\,|\dot{u}|^{\rho-1}\,\dot{u} & \text{if } \dot{u}\neq 0\\ 0 & \text{if } \dot{u}=0\end{cases}\qquad (8.75)$$

in the case of viscoelastic materials with long memory. Here, ρ is a regularization parameter and is assumed to satisfy $0 < \rho \leq 1$.

Equalities (8.70), (8.71), (8.73), and (8.74) are satisfied on $\Gamma_3 \times (0, T)$ and equalities (8.72) and (8.75) are satisfied on Γ_3, for every $t \in [0, T]$. In this book, we provide results in the study of antiplane contact problems involving friction laws of the form (8.70)–(8.72) under the assumption (8.68) on g. Part of these results can be extended to antiplane contact problems involving friction laws of the form (8.73)–(8.75). However, to avoid repetitions, we restrict ourselves to the case of elastic materials and, therefore, we use the friction law (8.73); we assume in this case that $g \in L^\infty(\Gamma_3)$ is a positive function.

The friction laws (8.60)–(8.62) model a situation in which surfaces are dry; they are characterized by the existence of a positive function, the friction bound, such that slip may occur only when the friction force reaches the critical value provided by the friction bound. The friction laws (8.70)–(8.75) do not have this feature, since in their case slip appears even for small tangential shear. Such kind of situations appear in practice when the contact surfaces are lubricated.

We note that the friction conditions above are formulated in the evolutionary case. The static case is obtained by replacing the velocity $\dot{u}$ with the displacement u in the corresponding formulas. To provide some examples, we see that the static version of (8.60) is given by

$$|\mu\,\partial_\nu u| \leq g, \quad \mu\,\partial_\nu u = -g\,\frac{u}{|u|} \quad \text{if } u \neq 0, \tag{8.76}$$

the static version of (8.70) is given by

$$\mu\,\partial_\nu\, u = -g\,\frac{u}{\sqrt{u^2 + \rho^2}}, \tag{8.77}$$

and the static version of (8.73) is given by

$$\mu\,\partial_\nu u = \begin{cases} -g\,|u|^{\rho-1}\,u & \text{if } u \neq 0 \\ 0 & \text{if } u = 0. \end{cases} \tag{8.78}$$

Equalities (8.76)–(8.78) are satisfied on Γ_3. Note also that the static version of the friction laws should be seen either as a model suitable for proportional loadings or as a first approximation of the evolutionary ones.

We end this section with the remark that in the physical setting of the antiplane problems described above, the normal component of the stress vanishes on the contact surface during the process, see (8.56). This feature is not incompatible with the choice (7.36) for the friction bound, which corresponds with the Tresca friction law. However, we note that the use of the choice (7.37) or (7.38) leads to the friction bound $g = 0$ and, therefore, the Coulomb law of dry friction (7.35) implies that the friction force vanishes on the contact surface, i.e., the contact is frictionless. We conclude that the choice (7.37) or (7.38) for the friction bound is not appropriate with the study of antiplane frictional problems in the physical setting above.

For this reason, in order to accommodate the classical and the modified Coulomb law of dry friction with the antiplane context, in the next section we present models for frictional contact that involve antiplane shear deformations in a slightly different physical setting. In these models, the normal stress on the contact surface does not vanish and, therefore, the use of the Coulomb law of dry friction with the choice (7.37) or (7.38) makes sense, as it is compatible with a genuine frictional condition.

8.5 Antiplane Models for Pre-stressed Cylinders

The models of antiplane shear deformation studied in the previous sections of this chapter were derived under the assumption that the reference configuration of the cylindrical body is *unstressed*, i.e., in absence of body forces and tractions, the stress tensor vanishes. However, there is a need to consider the case when the body is in a *pre-stressed* reference configuration, i.e., besides the external loads that give rise to stresses and deformations, the body is submitted to pre-loads that induce a primary stress. This stress, denoted in what follows $\boldsymbol{\sigma}^P$, plays the role of a "self-equilibrating stress," since it is in equilibrium with the pre-loads that keep the body in the reference configuration. Concrete situations of such type arise in the geophysical models that describe the systems of faults in a domain (see, e.g., [20, 74, 81]) as well as in the study of pre-stressed composite assemblies (see, e.g., [97, 118, 127]). They also arise in the study of mining engineering problems (see, e.g., [34, 35]), where the reference configuration of the rocks is supposed to be stressed before excavation.

In this section, we provide models for the antiplane frictional contact problems in the case when the cylindrical body B is in a pre-stressed reference configuration. To this end, we consider again the physical setting presented in Section 8.1 and conserve the notation introduced there; however, we assume now that there exists a primary stress $\boldsymbol{\sigma}^P$ in the reference configuration, obtained by applying pre-loads and, moreover,

$$\boldsymbol{\sigma}^P = \begin{pmatrix} -F & 0 & 0 \\ 0 & -F & 0 \\ 0 & 0 & 0 \end{pmatrix}, \tag{8.79}$$

where F is a positive constant. This last assumption is made only for the sake of simplicity, since by using the same arguments, the results below can be extended to the case when the primary stress has a more general form. Note also that the choice (8.79) of the primary stress is inspired by the geophysical context of modeling of earthquakes' evolution, where F is interpreted as the normal stress on the faults, see [19, 74, 76, 81] and the references therein.

It follows from assumption (8.79) that

$$\mathrm{Div}\, \boldsymbol{\sigma}^P = \mathbf{0} \tag{8.80}$$

and, using (7.28), (8.17), we obtain

$$\boldsymbol{\sigma}^P \boldsymbol{\nu} = -F\boldsymbol{\nu}, \quad \sigma_\nu^P = -F, \quad \boldsymbol{\sigma}_\tau^P = \mathbf{0}. \tag{8.81}$$

Equality (8.80) shows that the pre-loaded body force vanishes and (8.81) implies that the pre-loaded pressure on the cylinder's surface is $-F\boldsymbol{\nu}$.

Assume now the antiplane context (8.1)–(8.3) and denote by $\boldsymbol{\sigma} = (\sigma_{ij})$ the stress in the body associated to the deformation field (8.4), by using one of the constitutive laws (7.23), (7.24), or (7.25); then, by the arguments presented in Section 8.1, it follows that $\boldsymbol{\sigma}$ is of the form (8.12), where the components σ_{13} and σ_{23} depend only on x_1, x_2 and t; also, the arguments presented in Section 8.4 show that

$$\boldsymbol{\sigma}\boldsymbol{\nu} = (0, 0, q), \quad \sigma_\nu = 0, \quad \boldsymbol{\sigma}_\tau = (0, 0, q) \tag{8.82}$$

where $q = \sigma_{13}\nu_1 + \sigma_{23}\nu_2$. Next, the total stress in the body after applying the body forces and tractions, denoted $\widetilde{\boldsymbol{\sigma}}$, is given by

$$\widetilde{\boldsymbol{\sigma}} = \boldsymbol{\sigma} + \boldsymbol{\sigma}^P. \tag{8.83}$$

Therefore, combining (8.80)–(8.83) we find that

$$\mathrm{Div}\, \widetilde{\boldsymbol{\sigma}} = \mathrm{Div}\, \boldsymbol{\sigma}, \tag{8.84}$$

$$\widetilde{\boldsymbol{\sigma}}\boldsymbol{\nu} = \boldsymbol{\sigma}\boldsymbol{\nu} - F\boldsymbol{\nu}, \quad \widetilde{\sigma}_\nu = -F, \quad \widetilde{\boldsymbol{\sigma}}_\tau = \boldsymbol{\sigma}_\tau. \tag{8.85}$$

Since the pre-loaded body forces vanish, the equilibrium equation is given by

$$\mathrm{Div}\, \widetilde{\boldsymbol{\sigma}} + \boldsymbol{f}_0 = \mathbf{0} \quad \text{in } B \times (0, T)$$

and, using (8.84), we find that

$$\mathrm{Div}\, \boldsymbol{\sigma} + \boldsymbol{f}_0 = \mathbf{0} \quad \text{in } B \times (0, T).$$

We now use assumption (8.1) in the previous equality to recover equation (8.5). Then, we use the arguments in Section 8.1 to see that the equation of equilibrium is given by (8.7) in the elastic case (8.6), by (8.11) in the viscoelastic case with short memory (8.10), and by (8.15) in the case of viscoelastic materials with long memory (8.13), (8.14).

We now discuss the displacement-traction boundary conditions. Recall that since the cylinder is fixed on $\Gamma_D \times (0,T)$, the displacement field vanishes there and, therefore, by (7.2) and (8.3), we impose the displacement condition (8.16). Next, the traction boundary condition is

$$\widetilde{\boldsymbol{\sigma}}\boldsymbol{\nu} = \boldsymbol{f}_2 - F\boldsymbol{\nu} \quad \text{on } \Gamma_F \times (0,T),$$

and here we take into account the pre-loaded pressure $-F\boldsymbol{\nu}$ that is acting on the boundary Γ_F of the cylinder. Combining this last boundary condition with (8.85) we obtain

$$\boldsymbol{\sigma}\boldsymbol{\nu} = \boldsymbol{f}_2 \quad \text{on } \Gamma_F \times (0,T).$$

Then, by the arguments presented in Section 8.1, it follows that the traction boundary condition is given by (8.23) in the case of elastic materials, by (8.24) in the case of viscoelastic materials with short memory, and by (8.25) in the case of viscoelastic materials with long memory.

We now investigate the frictional contact conditions on $\Gamma_C \times (0,T)$. First, note that by (8.85) it follows that

$$\widetilde{\sigma}_\nu = -F \quad \text{on } \Gamma_C \times (0,T), \tag{8.86}$$

i.e., the normal stress is prescribed and does not vanish on the contact boundary during the process. This is the main difference with the case of unstressed configuration presented in Sections 8.1 and 8.4, as there the normal stress vanishes, see (8.56).

Assume now that the tangential part of the total stress $\widetilde{\boldsymbol{\sigma}}$ satisfies on $\Gamma_C \times (0,T)$ the Coulomb law of dry friction in its quasistatic form (7.35); since (8.85) implies that $\widetilde{\boldsymbol{\sigma}}_\tau = \boldsymbol{\sigma}_\tau$, it follows that the friction force $\boldsymbol{\sigma}_\tau$ satisfies the same friction law on $\Gamma_C \times (0,T)$. Therefore, arguing as in Section 8.4, we obtain that the friction conditions are formulated on $\Gamma_3 \times (0,T)$ and are given by (8.60) in the case of elastic materials, by (8.61) in the case of viscoelastic materials with short memory, and by (8.62) in the case of viscoelastic materials with long memory. Note also that now $\widetilde{\sigma}_\nu \neq 0$ and therefore it follows from the discussion on page 161 that, besides the Tresca friction law corresponding with the friction bound (7.36), it make sense to consider the classical Coulomb law of friction corresponding with the friction bound (7.37) as well as the modified Coulomb law of friction corresponding with the friction bound (7.38). Thus, as explained in Section 7.4, it makes sense to consider slip, slip rate, total slip, and total slip rate dependent friction laws, which in the antiplane context lead to friction bounds of the form (8.64)–(8.67).

Similar results can be obtained by considering the static version of the antiplane frictional processes. For instance, considering the static Coulomb law of dry friction, (7.60), we obtain that, in the case of elastic materials, the frictional contact conditions are given by (8.76).

From the discussion above, we conclude that considering antiplane frictional contact processes in the context of pre-stressed reference configurations in which the primary stress $\boldsymbol{\sigma}^P$ is of the form (8.79) leads to the equation of equilibrium and to the displacement-traction boundary conditions described in Section 8.1, as well as to the frictional conditions described in Section 8.4. The consequence of this conclusion is twofold. First, it shows that from the mathematical point of view, there is no difference between the study of antiplane frictional contact problems in unstressed reference configurations and the study of antiplane frictional contact problems in pre-stressed reference configurations, since these problems lead to the same mathematical models, even if their physical settings are different. Second, it justifies the use of the Coulomb law of dry friction with the friction bounds (8.64)–(8.67), since considering such friction bounds makes sense in the case of antiplane contact problems formulated on pre-stressed reference configurations.

For all these reasons, in Part IV of this book we use the equations and boundary conditions in Sections 8.1 and 8.4, without specifying if the reference configuration is unstressed or pre-stressed; we combine them to derive mathematical models in the study of antiplane frictional contact problems with elastic or viscoelastic materials; then, we provide the variational analysis of the models by using the abstract results presented in Part II of the book.

Bibliographical Notes

For a comprehensive treatment of basic aspects of solid mechanics, the reader is referred to [6, 40, 51, 56, 84, 94, 96, 109, 110, 143] and to [26] for an in-depth mathematical treatment of three-dimensional elasticity. More information concerning the elastic and viscoelastic constitutive laws presented in Section 7.2 can be found in [41, 116, 144]. One is referred to [82, 119] for more details on the modeling aspects of contact mechanics treated in the rest of Chapter 7. Experimental background and elements of surface physics that justify some of the contact and frictional boundary conditions presented in Sections 7.3 and 7.4 can be found in [55, 86, 104, 111, 112, 119, 146]. In particular, the relationship between the Coulomb and Tresca friction law is discussed in [119, 146] and, there, a natural transition from the Coulomb to the Tresca friction laws is justified. The regularization (7.50) of the Coulomb friction law was used in [75, 76] in the study of a dynamic contact problem with slip-dependent coefficient of friction involving viscoelastic and elastic materials, respectively. Its static version, (7.61), was used in [65]; there, the finite element uniqueness of a problem in linear elasticity with unilateral contact is investigated.

Contact problems in elasticity have been studied by many authors. The famous Signorini problem was formulated in [131] to model the unilateral contact between an elastic material and a rigid foundation. Mathematical analysis of the Signorini problem was provided in [47] and its numerical analysis was performed in [86]. References concerning the Signorini contact problem for elastic and inelastic materials include [142]. References treating modeling and analysis of various contact problems include [42, 43, 60, 80, 86, 113, 130, 149]. Many computational methods for problems in contact mechanics can be found in the monographs [93, 147] and in the extensive lists of references therein. The state of the art in the field can be found in the proceedings [98, 120, 148] and in the special issue [128], as well.

The model of antiplane shear deformation described in Chapter 8 of the book represents one of the simplest classes of deformations that a solid can undergo. For this reason, it was considered by many authors, and we mention below some of the references on this topic.

Following the pioneering works [48, 90, 91] that led to a modern revival of interest in antiplane shear deformations in hyperelasticity, several papers have been devoted to these deformations, see for instance [71, 72]. The purpose of [71] is to study the antiplane shear deformations of a nonhomogeneous anisotropic cylinder submitted to prescribed surface tractions on its lateral boundary; it is shown that, in the absence of body forces, not all arbitrary anisotropic cylinders will sustain an antiplane shear deformation; moreover, necessary and sufficient conditions on the elastic moduli are obtained that do allow an antiplane shear. The aim of [72] is to investigate the basic issues that arise when generalized plain strain deformations are superimposed on antiplane shear deformations in isotropic incompressible hyperelastic materials.

An excellent reference concerning antiplane shear deformations in solid mechanics is the review article [69], where modern developments for the antiplane shear model and its applications are described for both linear and nonlinear solid materials; various physical settings (including dynamic effects) where the antiplane shear model shows promise for further development are outlined; and some open questions concerning a challenging antiplane shear inverse problem in linear isotropic elastostatics are described.

Reference on antiplane problems in the finite strain framework include [21, 45, 62, 63, 145]. Thus, the equations governing antiplane shear motions for nonlinear elastic Mooney-Rivlin materials were first derived in [21] and extended in [63] to the case of viscoelastic Mooney-Rivlin materials. In [62], some exact dynamical solutions for antiplane problems involving neo-Hookean materials are determined. The equations for antiplane shear motions may be found in [145], in the general case of an anisotropic nonlinear elastic materials, and in [45], in the framework of the nonlinear theory of viscoelastic materials of differential type.

An antiplane problem for incompressible isotropic viscoplastic solids was considered in [54]; there, the authors use a flow rule that is a properly invariant generalization of Coulomb's sliding friction law and provide a uniqueness result in the study of the corresponding antiplane initial boundary value problem.

The model of antiplane shear was used in fracture mechanics to model the situation when the crack surfaces move relatively to one another and parallel to the leading edge of the crack, see for instance [64, p. 321] for details. Following [70], the damage and local crack tip process zone are modeled in fracture mechanics by using elastic antiplane problems in which the elastic coefficients vary continuous with position. References on this topic include [23, 114], where the cracks in viscoelastic materials are studied under the antiplane shear condition.

Part IV
Antiplane Frictional Contact Problems

Chapter 9
Elastic Problems

In this chapter, we study antiplane frictional contact models for elastic materials, both in the static and quasistatic case. We start by considering static problems in which friction is described with the Tresca law and its regularizations; we derive a variational formulation for each model that is in the form of an elliptic variational inequality for the displacement field; then we prove existence, uniqueness, and convergence results for the weak solution. Next, we consider a static slip dependent frictional contact problem that leads to an elliptic quasivariational inequality and, again, we prove the existence and uniqueness of the weak solution. We then extend part of the above results in the study of quasistatic versions of these models, which lead to evolutionary variational or quasivariational inequalities. Everywhere in this chapter, we use the space V (page 152) together with its inner product $(\cdot,\cdot)_V$ and the associated norm $\|\cdot\|_V$, and we denote by $[0,T]$ the time interval of interest, $T>0$. Also, everywhere in this chapter, the use of the abstract results presented in Part II of this manuscript is made in the case $X=V$, $(\cdot,\cdot)_X=(\cdot,\cdot)_V$, without explicit specification.

9.1 Static Frictional Problems

In this section, we consider static frictional antiplane problems. For the first problem, the friction is modeled with Tresca's law and therefore the classical model of the process is the following.

Problem 9.1. *Find a displacement field* $u:\Omega\to\mathbb{R}$ *such that*

$$\operatorname{div}(\mu\,\nabla u)+f_0=0 \qquad \text{in } \Omega, \tag{9.1}$$

$$u=0 \qquad \text{on } \Gamma_1, \tag{9.2}$$

$$\mu\,\partial_\nu u=f_2 \qquad \text{on } \Gamma_2, \tag{9.3}$$

$$|\mu\,\partial_\nu u|\le g,\quad \mu\,\partial_\nu u=-g\,\frac{u}{|u|}\quad \text{if } u\neq 0 \qquad \text{on } \Gamma_3. \tag{9.4}$$

Note that (9.1) represents the equation of equilibrium (8.7), (9.2) represents the displacement boundary condition (8.16), and (9.3) represents the traction boundary condition (8.23), in their static versions. Finally, (9.4) represent the static version of Tresca's law, (8.76).

We turn now to the variational formulation of Problem 9.1. To this end we assume in what follows that the body forces and surface tractions have the regularity

$$f_0 \in L^2(\Omega), \quad f_2 \in L^2(\Gamma_2). \tag{9.5}$$

We also assume that the Lamé coefficient and the friction bound satisfy

$$\begin{aligned} &\mu \in L^\infty(\Omega) \text{ and there exists } \mu^* > 0 \text{ such that} \\ &\qquad \mu(\boldsymbol{x}) \geq \mu^* \quad \text{a.e. } \boldsymbol{x} \in \Omega. \end{aligned} \tag{9.6}$$

$$g \in L^2(\Gamma_3) \;\; \text{and} \;\; g(\boldsymbol{x}) \geq 0 \quad \text{a.e. } \boldsymbol{x} \in \Gamma_3. \tag{9.7}$$

We define the bilinear form $a : V \times V \to \mathbb{R}$ and the functional $j : V \to \mathbb{R}$ by the equalities

$$a(u, v) = \int_\Omega \mu\, \nabla u \cdot \nabla v \, dx \quad \forall\, u,\, v \in V, \tag{9.8}$$

$$j(v) = \int_{\Gamma_3} g\, |v| \, da \quad \forall\, v \in V \tag{9.9}$$

and note that by assumptions (9.6)–(9.7), the integrals in (9.8)–(9.9) are well defined. We also use (9.5) and Riesz's representation theorem (page 11) to define the element $f \in V$ by the equality

$$(f, v)_V = \int_\Omega f_0\, v \, dx + \int_{\Gamma_2} f_2\, v \, da \quad \forall\, v \in V. \tag{9.10}$$

With these notations, we follow the standard procedure to derive the variational formulation of Problem 9.1. Some of the arguments we use were already described in Section 8.3 and, therefore, we skip the details.

Assume that the mechanical problem 9.1 has a solution u such that $u \in V$, $\mu\, \nabla u \in H^1(\Omega)^2$, and let $v \in V$. We multiply equation (9.1) with $v - u$, integrate the result on Ω, and use Green's formula (8.35) and notation (8.18) to find

$$\int_\Omega \mu\, \nabla u \cdot (\nabla v - \nabla u) \, dx = \int_\Omega f_0\, (v - u) \, dx + \int_\Gamma \mu\, \partial_\nu u\, (v - u) \, da.$$

We split the boundary integral over Γ_1, Γ_2, and Γ_3 and, since $v-u$ vanishes on Γ_1 and $\mu\,\partial_\nu u = f_2$ on Γ_2, we deduce that

$$\int_\Omega \mu\nabla u\cdot(\nabla v-\nabla u)\,dx$$
$$= \int_\Omega f_0\,(v-u)\,dx + \int_{\Gamma_2} f_2\,(v-u)\,da + \int_{\Gamma_3}\mu\,\partial_\nu u\,(v-u)\,da. \tag{9.11}$$

Now we use (9.4) to see that

$$\mu\,\partial_\nu u\,(v-u) \ge g\,|u| - g\,|v| \quad \text{a.e. on } \Gamma_3. \tag{9.12}$$

Indeed, in the points of Γ_3 where $u\neq 0$ we have

$$\mu\,\partial_\nu u\,(v-u) = -g\,\frac{u}{|u|}\,(v-u) = g|u| - g\,\frac{uv}{|u|} \ge g|u| - g|v|,$$

as $\frac{uv}{|u|}\le |v|$; in the points of Γ_3 where $u=0$ we have

$$\mu\,\partial_\nu u\,(v-u) = \mu\,(\partial_\nu u)\,v \ge -|\mu\,\partial_\nu u|\,|v| \ge -g|v| = g\,|u| - g\,|v|,$$

as $|\mu\,\partial_\nu u|\le g$.

We integrate (9.12) on Γ_3 to obtain

$$\int_{\Gamma_3}\mu\,\partial_\nu u\,(v-u)\,da \ge \int_{\Gamma_3} g\,|u|\,da - \int_{\Gamma_3} g\,|v|\,da$$

and combine this inequality with (9.11) to find

$$\int_\Omega \mu\,\nabla u\cdot(\nabla v-\nabla u)\,dx + \int_{\Gamma_3} g\,|v|\,da - \int_{\Gamma_3} g\,|u|\,da$$
$$\ge \int_\Omega f_0\,(v-u)\,dx + \int_{\Gamma_2} f_2\,(v-u)\,da. \tag{9.13}$$

We use now the notation (9.8)–(9.10) and inequality (9.13) to derive the following variational formulation of the antiplane frictional contact problem 9.1.

Problem 9.2. *Find a displacement field $u\in V$ such that*

$$a(u,v-u) + j(v) - j(u) \ge (f,v-u)_V \quad \forall\, v\in V. \tag{9.14}$$

In the study of Problem 9.2 we have the following existence, uniqueness, and continuous dependence result.

Theorem 9.3. *Assume that* (9.5)–(9.7) *hold. Then Problem* 9.2 *has a unique solution $u\in V$. Moreover, the mapping $f\mapsto u{:}V\to V$ is Lipschitz continuous.*

Proof. The bilinear form a defined by (9.8) is symmetric, and using (8.51) and (8.52) it follows that it is continuous and V-elliptic. We use (9.7) to see that the functional j defined by (9.9) is a seminorm on V and it satisfies

$$j(v) \le \|g\|_{L^2(\Gamma_3)} \|v\|_{L^2(\Gamma_3)} \quad \forall\, v \in V.$$

Therefore, (8.33) yields

$$j(v) \le c_0\, \|g\|_{L^2(\Gamma_3)}\, \|v\|_V \quad \forall\, v \in V. \tag{9.15}$$

It follows from (9.15) that the seminorm j is continuous, which implies that it is a convex lower semicontinuous function on V. Theorem 9.3 is now a consequence of Theorem 3.1. □

We conclude by Theorem 9.3 that the mechanical problem 9.1 has a unique weak solution that depends Lipschitz continuously on the data. Moreover, as it follows from the proof of Theorem 3.1, the solution minimizes on V the energy functional J defined by

$$J(v) = \frac{1}{2}\int_{\Omega} \mu\, \|\nabla v\|^2\, dx + \int_{\Gamma_3} g\, |v|\, da - \int_{\Omega} f_0\, v\, dx - \int_{\Gamma_2} f_2\, v\, da.$$

We now study the problem in the case when the Tresca friction law (9.4) is replaced by its regularized versions. We start by considering the following problem.

Problem 9.4. *Find a displacement field $u_\rho : \Omega \to \mathbb{R}$ such that*

$$\operatorname{div}(\mu\, \nabla u_\rho) + f_0 = 0 \quad \text{in } \Omega, \tag{9.16}$$

$$u_\rho = 0 \quad \text{on } \Gamma_1, \tag{9.17}$$

$$\mu\, \partial_\nu u_\rho = f_2 \quad \text{on } \Gamma_2, \tag{9.18}$$

$$\mu\, \partial_\nu u_\rho = -g\, \frac{u_\rho}{\sqrt{u_\rho^2 + \rho^2}} \quad \text{on } \Gamma_3. \tag{9.19}$$

The equation and boundary conditions involved in Problem 9.4 have the same meaning as those in Problem 9.1. The difference arises from the fact that here we replace the static version of Tresca law (9.4) with its regularization (9.19), introduced in (8.77), page 161. Recall that here $\rho > 0$ is a regularization parameter.

The variational formulation of the antiplane frictional contact problem 9.4 is derived by using arguments similar to those used to derive the variational formulation of Problem 9.1. It is obtained in several steps that are described below.

Let $v \in V$ and assume that u_ρ is a regular solution to Problem 9.4. First, we prove that

$$\int_\Omega \mu \nabla u_\rho \cdot (\nabla v - \nabla u_\rho)\, dx = \int_\Omega f_0\, (v - u_\rho)\, dx + \int_{\Gamma_2} f_2\, (v - u_\rho)\, da + \int_{\Gamma_3} \mu\, \partial_\nu u_\rho\, (v - u_\rho)\, da. \tag{9.20}$$

Next, using (9.19) we easily deduce that

$$\mu\, \partial_\nu u_\rho\, (v - u_\rho) \geq g\sqrt{u_\rho^2 + \rho^2} - g\sqrt{v^2 + \rho^2} \quad \text{a.e. on } \Gamma_3. \tag{9.21}$$

We integrate now (9.21) on Γ_3 to obtain

$$\int_{\Gamma_3} \mu\, \partial_\nu u_\rho\, (v - u_\rho)\, da \geq \int_{\Gamma_3} g\sqrt{u_\rho^2 + \rho^2}\, da - \int_{\Gamma_3} g\sqrt{v^2 + \rho^2}\, da$$

and combine this last inequality with (9.20) to find

$$\int_\Omega \mu \nabla u_\rho \cdot (\nabla v - \nabla u_\rho)\, dx + \int_{\Gamma_3} \Big(g\sqrt{v^2 + \rho^2} - \rho\Big) da - \int_{\Gamma_3} \Big(g\sqrt{u_\rho^2 + \rho^2} - \rho\Big)\, da \geq \int_\Omega f_0\, (v - u_\rho)\, dx + \int_{\Gamma_2} f_2\, (v - u_\rho)\, da. \tag{9.22}$$

Finally, we define the functional $j_\rho : V \to \mathbb{R}$ by

$$j_\rho(v) = \int_{\Gamma_3} g\Big(\sqrt{v^2 + \rho^2} - \rho\Big)\, da \quad \forall\, v \in V, \tag{9.23}$$

then we use notation (9.8), (9.10), (9.23), and inequality (9.22) to derive the following variational formulation of the antiplane frictional contact problem 9.4.

Problem 9.5. *Find a displacement field $u_\rho \in V$ such that*

$$a(u_\rho, v - u_\rho) + j_\rho(v) - j_\rho(u_\rho) \geq (f, v - u_\rho)_V \quad \forall\, v \in V. \tag{9.24}$$

We have the following result.

Theorem 9.6. *Assume that* (9.5)–(9.7) *hold. Then, for every $\rho > 0$, Problem* 9.5 *has a unique solution $u_\rho \in V$. Moreover, the solution converges to the solution u of Problem* 9.2, *i.e.,*

$$u_\rho \to u \quad \text{in } V \text{ as } \rho \to 0. \tag{9.25}$$

Some of the arguments we need in the proof of Theorem 9.6 will be used several times in the rest of the book and for this reason we present them in the following lemma.

Lemma 9.7. *Assume that* (9.7) *holds. Then:*

(1) For every $\rho > 0$ *the functional* j_ρ*, defined by* (9.23)*, satisfies condition* (5.4) *on the space* $X = V$*.*

(2) The functionals j_ρ *and* j *satisfy condition* (3.17) *on* $X = V$*.*

Proof. (1) Let $\rho > 0$ and note that j_ρ is a positive convex functional that vanishes for $v = 0_V$. Moreover, for all $u,\, v \in V$ and $t \neq 0$ we have

$$\frac{j_\rho(u+tv) - j_\rho(u)}{t} = \int_{\Gamma_3} g\, \frac{\sqrt{(u+tv)^2+\rho^2} - \sqrt{u^2+\rho^2}}{t}\, da. \tag{9.26}$$

We also note that, as $t \to 0$, the convergence below holds:

$$g\, \frac{\sqrt{(u+tv)^2+\rho^2} - \sqrt{u^2+\rho^2}}{t} \to g\, \frac{uv}{\sqrt{u^2+\rho^2}} \quad \text{a.e. on } \Gamma_3. \tag{9.27}$$

Also, for all $u,\, v \in V$ and $t \neq 0$, the following inequalities are valid:

$$\begin{aligned}
&\left| g\, \frac{\sqrt{(u+tv)^2+\rho^2} - \sqrt{u^2+\rho^2}}{t} \right| \\
&\quad = \left| g\, \frac{2uv + tv^2}{\sqrt{(u+tv)^2+\rho^2} + \sqrt{u^2+\rho^2}} \right| \\
&\quad \le g\, \frac{|u+tv|\,|v|}{\sqrt{(u+tv)^2+\rho^2} + \sqrt{u^2+\rho^2}} + g\, \frac{|u|\,|v|}{\sqrt{(u+tv)^2+\rho^2} + \sqrt{u^2+\rho^2}} \\
&\quad \le g\, \frac{|u+tv|\,|v|}{\sqrt{(u+tv)^2+\rho^2}} + g\, \frac{|u|\,|v|}{\sqrt{u^2+\rho^2}} \quad \text{a.e. on } \Gamma_3.
\end{aligned}$$

Using now the inequality

$$\frac{|x|}{\sqrt{x^2+\rho^2}} \le 1$$

yields

$$\left| g\, \frac{\sqrt{(u+tv)^2+\rho^2} - \sqrt{u^2+\rho^2}}{t} \right| \le 2g\,|v| \quad \text{a.e. on } \Gamma_3. \tag{9.28}$$

We use (9.26)–(9.28) and Lebesgue's theorem to deduce that

$$\lim_{t\to 0} \frac{j_\rho(u+tv) - j_\rho(u)}{t} = \int_{\Gamma_3} g\, \frac{uv}{\sqrt{u^2+\rho^2}}\, da \quad \forall\, u,\, v \in V. \tag{9.29}$$

Since

$$v \mapsto \int_{\Gamma_3} g\, \frac{uv}{\sqrt{u^2+\rho^2}}\, da$$

is a linear continuous functional on V, from (9.29) and Defintition 1.35 we deduce that j_ρ is Gâteaux differentiable and, moreover,

$$(\nabla j_\rho(u), v)_V = \int_{\Gamma_3} g \, \frac{uv}{\sqrt{u^2+\rho^2}} \, da \quad \forall\, u,\, v \in V.$$

It follows from this equality that

$$\begin{aligned}(\nabla j_\rho(u), v)_V &\le \int_{\Gamma_3} g \, \frac{|u|\,|v|}{\sqrt{u^2+\rho^2}} \, da \\ &\le \int_{\Gamma_3} g\,|v|\, da \le \|g\|_{L^2(\Gamma_3)} \|v\|_{L^2(\Gamma_3)} \quad \forall\, u,\, v \in V\end{aligned}$$

and, using (8.33), we obtain

$$(\nabla j_\rho(u), v)_V \le c_0 \, \|g\|_{L^2(\Gamma_3)} \|v\|_V \quad \forall\, u,\, v \in V.$$

We choose $v = \nabla j_\rho(u)$ in the previous inequality to find

$$\|\nabla j_\rho(u)\|_X \le c_0 \, \|g\|_{L^2(\Gamma_3)} \quad \forall\, u \in V,$$

which shows that (5.4)(c) holds, and this concludes the first part of the lemma.

(2) Let $\rho > 0$. We note that for all $v \in V$, the following inequality holds,

$$\left|\sqrt{v^2+\rho^2} - \rho - |v|\right| = \rho + |v| - \sqrt{v^2+\rho^2} \le \rho \quad \text{a.e. on } \Gamma_3,$$

which implies that

$$\int_{\Gamma_3} g \left|\sqrt{v^2+\rho^2} - \rho - |v|\right| da \le \rho \int_{\Gamma_3} g \, da. \tag{9.30}$$

On the other hand, by the definition of the functionals j_ρ and j we obtain

$$|j_\rho(v) - j(v)| \le \int_{\Gamma_3} g \left|\sqrt{v^2+\rho^2} - \rho - |v|\right| da. \tag{9.31}$$

We combine (9.30) and (9.31) to find that

$$|j_\rho(v) - j(v)| \le \rho \int_{\Gamma_3} g \, da \quad \forall\, v \in V. \tag{9.32}$$

We deduce from (9.32) and (9.7) that j_ρ and j satisfy condition (3.17) on V with the choice $F(\rho) = \rho \displaystyle\int_{\Gamma_3} g \, da \ge 0$. □

We now prove Theorem 9.6.

Proof. Let $\rho > 0$. The unique solvability of Problem 9.5 follows again from Theorem 3.1, as by Lemma 9.7(1) and Corollary 1.37 it follows that the functional j_ρ defined by (9.23) is convex and l.s.c. Also, the convergence (9.25) is a consequence of Theorem 3.6 and Lemma 9.7(2). $\Box$

We now present similar results in the case of the power-law friction. In this case, the classical model of the process is the following.

Problem 9.8. *Find a displacement field* $u_\rho : \Omega \to \mathbb{R}$ *such that*

$$\operatorname{div}(\mu \nabla u_\rho) + f_0 = 0 \qquad \text{in } \Omega, \tag{9.33}$$

$$u_\rho = 0 \qquad \text{on } \Gamma_1, \tag{9.34}$$

$$\mu\, \partial_\nu u_\rho = f_2 \qquad \text{on } \Gamma_2, \tag{9.35}$$

$$\mu\, \partial_\nu u_\rho = \begin{cases} -g\,|u_\rho|^{\rho-1}\, u_\rho & \text{if } u_\rho \neq 0 \\ 0 & \text{if } u_\rho = 0 \end{cases} \qquad \text{on } \Gamma_3. \tag{9.36}$$

The equations and boundary conditions involved in Problem 9.8 have the same meaning as those involved in Problem 9.1. The difference arises from the fact that here we replace the static version of the Tresca law (9.4) with the static version of the power-law friction (9.36), already introduced by formula (8.78), page 161. Recall that here ρ is a regularization parameter and is assumed to satisfy $0 < \rho \leq 1$.

In the study of Problem 9.8, we keep assumptions (9.5), (9.6) and we strengthen assumption (9.7) as follows:

$$g \in L^\infty(\Gamma_3) \ \text{ and } \ g(\boldsymbol{x}) \geq 0 \quad \text{ a.e. } \boldsymbol{x} \in \Gamma_3. \tag{9.37}$$

The variational formulation of the antiplane frictional problem 9.8 is based on similar arguments as above. Note that in this case inequality (9.21) is replaced by the inequality

$$\mu\, \partial_\nu u_\rho\, (v - u_\rho) \geq \frac{g}{\rho+1}\, |u_\rho|^{\rho+1} - \frac{g}{\rho+1}\, |v|^{\rho+1} \quad \text{ a.e. on } \Gamma_3, \tag{9.38}$$

which follows from (9.36) and inequality (1.21) derived in Lemma 1.46. As a consequence, (9.22) is replaced by the inequality

$$\int_\Omega \mu \nabla u_\rho \cdot (\nabla v - \nabla u_\rho) dx + \frac{1}{\rho+1} \int_{\Gamma_3} g\, |v|^{\rho+1}\, da - \frac{1}{\rho+1} \int_{\Gamma_3} g\, |u_\rho|^{\rho+1}\, da$$
$$\geq \int_\Omega f_0\, (v - u_\rho)\, dx + \int_{\Gamma_2} f_2\, (v - u_\rho)\, da. \tag{9.39}$$

Below in this section, we denote by $j_\rho : V \to \mathbb{R}$ the functional given by

$$j_\rho(v) = \frac{1}{\rho+1} \int_{\Gamma_3} g\, |v|^{\rho+1}\, da \quad \forall\, v \in V \tag{9.40}$$

and we note that the integral above is well defined, as $0 < \rho \leq 1$ and g satisfies (9.37). We use notation (9.8), (9.10), (9.40), and inequality (9.39) to obtain the following variational formulation of the antiplane frictional contact problem 9.8.

Problem 9.9. *Find a displacement field* $u_\rho \in V$ *such that*

$$a(u_\rho, v - u_\rho) + j_\rho(v) - j_\rho(u_\rho) \geq (f, v - u_\rho)_V \quad \forall\, v \in V. \tag{9.41}$$

We have the following result.

Theorem 9.10. *Assume that* (9.5), (9.6), *and* (9.37) *hold. Then, for every* $\rho \in (0,1]$, *Problem* 9.9 *has a unique solution* $u_\rho \in V$. *Moreover, the solution converges to the solution* u *of Problem* 9.2, *i.e.,*

$$u_\rho \to u \quad \text{in } V \text{ as } \rho \to 0. \tag{9.42}$$

Proof. Let $\rho \in (0,1]$ and let $\phi_\rho : \mathbb{R} \to \mathbb{R}$ be the function defined by

$$\phi_\rho(v) = \begin{cases} |v|^{\rho-1}\, v & \text{if } v \neq 0 \\ 0 & \text{if } v = 0. \end{cases}$$

We use again Lebesgue's convergence theorem to see that

$$\lim_{t \to 0} \frac{j_\rho(u + tv) - j_\rho(u)}{t} = \int_{\Gamma_3} g\, \phi_\rho(u) v\, da \quad \forall\, u,\, v \in V;$$

therefore, since

$$v \mapsto \int_{\Gamma_3} g\, \phi_\rho(u) v\, da$$

is a linear continuous functional on V, we deduce that j_ρ is Gâteaux differentiable and, moreover,

$$(\nabla j_\rho(u), v)_V = \int_{\Gamma_3} g\, \phi_\rho(u) v\, da \quad \forall\, u,\, v \in V.$$

The functional j_ρ defined by (9.40) is convex and it follows from Corollary 1.37 that it is lower semicontinuous. Therefore, the existence and uniqueness part in Theorem 9.10 is a direct consequence of Theorem 3.1.

We turn now to the proof of the convergence (9.42) and, to this end, we show that j_ρ satisfies condition (3.18) on the space $X = V$. Note that (9.37) implies that $j_\rho(v) \geq 0\ \forall\, v \in V$ and, since $j_\rho(0_V) = 0$, we conclude that j_ρ satisfies condition (3.18)(a). Also, for all $v \in V$ we have

$$\frac{1}{\rho + 1}\, g\, |v|^{\rho+1} \to g\, |v| \quad \text{a.e. on } \Gamma_3 \quad \text{as } \rho \to 0,$$
$$\frac{1}{\rho + 1}\, g\, |v|^{\rho+1} \leq g\, (|v|^2 + 1) \quad \text{a.e. on } \Gamma_3 \quad \forall\, \rho \in (0,1].$$

Therefore, using again Lebesgue's convergence theorem, it follows that

$$\frac{1}{\rho+1}\int_{\Gamma_3} g\,|v|^{\rho+1}\,da \to \int_{\Gamma_3} g\,|v| \quad \forall\, v\in V,$$

which shows that (3.18)(b) holds.

Finally, to prove (3.18)(c), we consider a sequence $\{v_\rho\}\subset V$ such that $v_\rho \rightharpoonup v\in V$ as $\rho\to 0$ and we use inequality (1.21) to obtain

$$\begin{aligned} j_\rho(v_\rho)-j_\rho(v) &= \frac{1}{\rho+1}\int_{\Gamma_3} g\,|v_\rho|^{\rho+1}\,da - \frac{1}{\rho+1}\int_{\Gamma_3} g\,|v|^{\rho+1}\,da \\ &\geq \int_{\Gamma_3} g\,\phi_\rho(v)(v_\rho - v)\,da. \end{aligned} \tag{9.43}$$

Next, since $|\phi_\rho(v)|\leq |v|+1$ a.e. on Γ_3, we obtain that

$$\begin{aligned} \left|\int_{\Gamma_3} g\,\phi_\rho(v)(v_\rho - v)\,da\right| &\leq \int_{\Gamma_3} g\,|\phi_\rho(v)|\,|v_\rho - v|\,da \\ &\leq \int_{\Gamma_3} g\,(|v|+1)\,|\,|v_\rho - v|\,da. \end{aligned}$$

Also, we note that the weak convergence $v_\rho \rightharpoonup v$ in V and the compactness of the trace operator imply the strong convergence $v_\rho \to v$ in $L^2(\Gamma_3)$ and, therefore, from the previous inequality we obtain that

$$\lim_{\rho\to 0}\int_{\Gamma_3} g\,\phi_\rho(v)(v_\rho - v)\,da = 0. \tag{9.44}$$

We now combine (9.43) and (9.44) to see that

$$\liminf_{\rho\to 0}\big(j_\rho(v_\rho)-j_\rho(v)\big)\geq 0 \tag{9.45}$$

and, on the other hand, recall that (3.18)(b) implies that

$$\lim_{\rho\to 0} j_\rho(v) = j(v). \tag{9.46}$$

We now combine (9.45) and (9.46) and find

$$\liminf_{\rho\to 0} j_\rho(v_\rho) = \liminf_{\rho\to 0}\big(j_\rho(v_\rho)-j_\rho(v)+j_\rho(v)\big)\geq j(v)$$

which shows that j_ρ satisfies (3.18)(c). The convergence (9.42) is now a consequence of Theorem 3.6, which concludes the proof. □

We conclude by Theorems 9.6 and 9.10 that Problems 9.4 and 9.8 have a unique weak solution that converges to the weak solution of Problem 9.1 as $\rho\to 0$. In addition to the mathematical interest in this result, it is important from the mechanical point of view since it shows that, in the study of static antiplane contact problems for elastic materials, the solution of the problem with Tresca's friction law can be approached as closely as one wishes by the solution of the problem with regularized friction law or with power-law friction, with a sufficiently small parameter ρ. Moreover, since the functionals j_ρ defined in (9.23) and (9.40) are Gâteaux differentiable, by using the arguments presented on page 50, it follows that solving the variational inequality (9.24) or (9.41) reduces to the solution of a nonlinear equation on the space V, which is more convenient from the numerical point of view.

9.2 A Static Slip-dependent Frictional Problem

For the problem considered in this section, the friction condition is described with the static version of the slip-dependent Coulomb's law. This law is of the form (8.76) in which the friction bound depends on $|u|$, i.e., $g = g(|u|)$. Therefore, the classical formulation of the problem is as follows.

Problem 9.11. *Find a displacement field $u : \Omega \to \mathbb{R}$ such that*

$$\operatorname{div}(\mu \nabla u) + f_0 = 0 \qquad \text{in } \Omega, \tag{9.47}$$

$$u = 0 \qquad \text{on } \Gamma_1, \tag{9.48}$$

$$\mu\, \partial_\nu u = f_2 \qquad \text{on } \Gamma_2, \tag{9.49}$$

$$|\mu\, \partial_\nu u| \le g(|u|), \quad \mu\, \partial_\nu u = -g(|u|)\, \frac{u}{|u|} \quad \text{if } u \neq 0 \quad \text{on } \Gamma_3. \tag{9.50}$$

In the study of this problem, we assume that (9.5), (9.6), and (8.69) hold. We use the bilinear form a given by (9.8), the element f given by (9.10), and the functional $j : V \times V \to \mathbb{R}$ given by

$$j(u, v) = \int_{\Gamma_3} g(|u|)\, |v|\, da \quad \forall\, u, v \in V. \tag{9.51}$$

Note that assumption (8.69) on the friction bound implies that for all $v \in V$, the function $\boldsymbol{x} \mapsto g(\boldsymbol{x}, |v(\boldsymbol{x})|)$ belongs to $L^2(\Gamma_3)$ and therefore the integral in (9.51) is well defined.

Using arguments similar to those presented in Section 9.1, we derive the following variational formulation of Problem 9.11.

Problem 9.12. *Find a displacement field $u \in V$ such that*

$$a(u, v - u) + j(u, v) - j(u, u) \ge (f, v - u)_V \quad \forall\, v \in V. \tag{9.52}$$

We conclude that the variational formulation of the elastic slip-dependent antiplane frictional problem is given by an elliptic quasivariational inequality and, therefore, we turn to the results presented in Section 3.3 to solve this problem.

Our first result in the study of Problem 9.12 is the following.

Theorem 9.13. *Assume that* (9.5), (9.6), *and* (8.69) *hold. Then there exists a constant* L_0*, which depends only on* Ω, Γ_1, Γ_3, *and* μ, *such that problem* (9.52) *has a unique solution* u *that depends Lipschitz continuously on* f, *if* $L_g < L_0$.

Proof. For all $\eta \in V$, the functional $j(\eta, \cdot) : V \to \mathbb{R}$ is a continuous seminorm and therefore is convex and l.s.c., i.e., it satisfies (3.30)(a). Let u_1, u_2, v_1, $v_2 \in V$; by using (9.51) and (8.69), we find that

$$\begin{aligned} & j(u_1, v_2) - j(u_1, v_1) + j(u_2, v_1) - j(u_2, v_2) \\ & = \int_{\Gamma_3} \Big(g(|u_1|) - g(|u_2|) \Big) (|v_2| - |v_1|) \, da \\ & \leq \int_{\Gamma_3} L_g |u_1 - u_2| \, |v_1 - v_2| \, da \\ & \leq L_g \|u_1 - u_2\|_{L^2(\Gamma_3)} \, \|v_1 - v_2\|_{L^2(\Gamma_3)} \end{aligned}$$

and, keeping in mind (8.33), we obtain

$$\begin{aligned} & j(u_1, \, v_2) - j(u_1, \, v_1) + j(u_2, \, v_1) - j(u_2, \, v_2) \\ & \quad \leq c_0^2 \, L_g \, \|u_1 - u_2\|_V \, \|v_1 - v_2\|_V. \end{aligned} \tag{9.53}$$

Let

$$L_0 = \frac{\mu^*}{c_0^2}. \tag{9.54}$$

Clearly, L_0 depends only on Ω, Γ_1, Γ_3, and μ. It follows from (9.53) that j satisfies condition (3.30)(b) with $\alpha = c_0^2 \, L_g$ and, moreover, it follows from (8.52) that a satisfies condition (3.2) with $m = \mu^*$. Assume now that $L_g < L_0$; then we obtain $c_0^2 \, L_g < \mu^*$, which implies that $m > \alpha$. Theorem 9.13 follows now from Theorem 3.7 on page 51. □

We conclude by Theorem 9.13 that, under a smallness assumption on L_g, Problem 9.12 has a unique weak solution that depends Lipschitz continuously on the data.

We now prove the existence of a solution to the quasivariational inequality (9.52) without any smallness assumption on L_g.

Theorem 9.14. *Assume that* (9.5), (9.6), *and* (8.69) *hold. Then there exists at least one solution to Problem* 9.12.

Proof. We use Theorem 3.10. To this end, we note that the functional j given by (9.51) satisfies condition (3.36). Let us now consider the sequences $\{\eta_n\} \subset V$, $\{u_n\} \subset V$ such that $\eta_n \rightharpoonup \eta \in V$ and $u_n \rightharpoonup u \in V$. Using the compactness property of the trace operator and (8.69), we obtain that

$$g(|\eta_n|) \to g(|\eta|) \quad \text{in } L^2(\Gamma_3),$$
$$|u_n| \to |u| \quad \text{in } L^2(\Gamma_3).$$

Therefore, by the definition of the functional j, we deduce that

$$j(\eta_n, v) \to j(\eta, v) \quad \forall\, v \in V,$$
$$j(\eta_n, u_n) \to j(\eta, u),$$

which show that the functional j satisfies condition (3.37). Finally, recall that V is a separable Hilbert space, as proved in Theorem 8.1. Theorem 9.14 follows now from Theorem 3.10 on page 53. □

Note that Theorem 3.10 can be used to prove the uniqueness of the solution of Problem 9.12 and its Lipschitz continuous dependence on the data under the smallness assumption in Theorem 9.13. Indeed, we choose $u_1 = v_1 = u$ and $u_2 = v_2 = v$ in (9.53) to obtain that

$$j(u, v) - j(u, u) + j(v, u) - j(u, u) \le c_0^2\, L_g\, \|u - v\|_V^2$$

and, choosing $L_g < L_0$ with L_0 given by (9.54), we deduce that j satisfies condition (3.39). We conclude by Theorem 3.10 (3) that, if $L_g < L_0$, then the solution of Problem 9.12 is unique and depends Lipschitz continuously on f.

9.3 Quasistatic Frictional Problems

In this section, we study the quasistatic versions of the first two frictional contact problems presented in Section 9.1. We apply the results presented in Sections 5.1 and 5.2 to prove the existence of a unique weak solution to each one of these problems.

To construct the quasistatic version of Problem 9.1, we consider a time interval of interest $[0, T]$ with $T > 0$, allow the body forces and tractions to depend on time, replace (9.4) with the evolutionary version of Tresca's friction law, (8.60), and add an initial condition for the displacement field. Therefore, the problem can be formulated as follows.

Problem 9.15. *Find a displacement field $u : \Omega \times [0, T] \to \mathbb{R}$ such that*

$$\mathrm{div}\,(\mu\, \nabla u) + f_0 = 0 \qquad \text{in } \Omega \times (0, T), \tag{9.55}$$

$$u = 0 \qquad \text{on } \Gamma_1 \times (0, T), \tag{9.56}$$

$$\mu\, \partial_\nu u = f_2 \qquad \text{on } \Gamma_2 \times (0, T), \tag{9.57}$$

$$|\mu\,\partial_\nu u| \le g, \quad \mu\,\partial_\nu u = -g\,\frac{\dot{u}}{|\dot{u}|} \quad \text{if } \dot{u} \ne 0 \quad \text{on } \Gamma_3 \times (0,T), \tag{9.58}$$

$$u(0) = u_0 \qquad \text{in } \Omega. \tag{9.59}$$

In the study of this problem, we assume that the Lamé coefficient μ and the friction bound g satisfy condition (9.6) and (9.7), respectively; moreover, we assume that the body forces and surface tractions satisfy

$$f_0 \in W^{1,2}(0,T;L^2(\Omega)), \quad f_2 \in W^{1,2}(0,T;L^2(\Gamma_2)). \tag{9.60}$$

We use notation (9.8) and (9.9) for the bilinear form a and the functional j; also, we use Riesz's representation theorem to define the function $f : [0,T] \to V$ by the equality

$$(f(t),\, v)_V = \int_\Omega f_0(t)\, v\, dx + \int_{\Gamma_2} f_2(t)\, v\, da \quad \forall\, v \in V,\ t \in [0,T]. \tag{9.61}$$

It follows from (9.60) that the integrals in (9.61) are well defined and, moreover,

$$f \in W^{1,2}(0,T;V). \tag{9.62}$$

Finally, we assume that the initial displacement u_0 satisfies

$$u_0 \in V, \tag{9.63}$$

$$a(u_0,\, v) + j(v) \ge (f(0),\, v)_V \quad \forall\, v \in V. \tag{9.64}$$

Condition (9.64) represents a compatibility condition on the initial data that is necessary in many quasistatic problems, see for instance [130]. Physically, it is needed so as to guarantee that initially the state is in equilibrium, since otherwise the inertial terms cannot be neglected and the problems become dynamic.

Proceeding as in the case of the static problem 9.1, we obtain the following variational formulation of the contact problem (9.55)–(9.59).

Problem 9.16. *Find a displacement field $u : [0,T] \to V$ such that*

$$\begin{aligned} &a(u(t), v - \dot{u}(t)) + j(v) - j(\dot{u}(t)) \\ &\quad \ge (f(t), v - \dot{u}(t))_V \quad \forall\, v \in V, \ \text{a.e. } t \in (0,T), \end{aligned} \tag{9.65}$$

$$u(0) = u_0. \tag{9.66}$$

We conclude that the variational formulation of the quasistatic problem 9.15 is given by an evolutionary variational inequality and, therefore, we prove the unique solvability of Problem 9.16 by using the regularization method presented in Section 5.2. The interest in this method arises from the fact that it is constructive, i.e., it provides the existence of the solution by passing to the limit in a sequence of regularized problems. We start by considering the following problem, for $\rho > 0$.

Problem 9.17. *Find a displacement field* $u_\rho : \Omega \times [0,T] \to \mathbb{R}$ *such that*

$$\operatorname{div}(\mu \nabla u_\rho) + f_0 = 0 \qquad \text{in } \Omega \times (0,T), \tag{9.67}$$

$$u_\rho = 0 \qquad \text{on } \Gamma_1 \times (0,T), \tag{9.68}$$

$$\mu\, \partial_\nu u_\rho = f_2 \qquad \text{on } \Gamma_2 \times (0,T), \tag{9.69}$$

$$\mu\, \partial_\nu u_\rho = -g \frac{\dot{u}_\rho}{\sqrt{\dot{u}_\rho^2 + \rho^2}} \qquad \text{on } \Gamma_3 \times (0,T), \tag{9.70}$$

$$u_\rho(0) = u_0 \qquad \text{in } \Omega. \tag{9.71}$$

Note that Problem 9.17 represents the quasistatic version of Problem 9.4, page 174. Also, it represents a regularization of Problem 9.15 since it is obtained from this last problem by replacing the Tresca friction law (9.58) with its regularization (9.70).

In the study of this problem, we assume that (9.6), (9.7), (9.60), (9.63), and (9.64) hold. Proceeding as in the case of the static problem 9.4, by using notation (9.8), (9.23), and (9.61) we obtain the following variational formulation of the contact problem (9.67)–(9.71).

Problem 9.18. *Find a displacement field* $u_\rho : [0,T] \to V$ *such that*

$$\begin{aligned} a(u_\rho(t), v - \dot{u}_\rho(t)) &+ j_\rho(v) - j_\rho(\dot{u}_\rho(t)) \\ &\geq (f(t), v - \dot{u}_\rho(t))_V \quad \forall\, v \in V, \quad \text{a.e. } t \in (0,T), \end{aligned} \tag{9.72}$$

$$u(0) = u_0. \tag{9.73}$$

We have the following existence, uniqueness, and convergence result.

Theorem 9.19. *Assume conditions* (9.6), (9.7), (9.60), (9.63), *and* (9.64). *Then:*

(1) For each $\rho > 0$, *there exists a unique solution to Problem* 9.18 *and it satisfies* $u_\rho \in W^{1,2}(0,T;V)$.

(2) There exists a unique solution to Problem 9.16 *and it satisfies* $u \in W^{1,2}(0,T;V)$.

(3) The solution u_ρ *of Problem* 9.18 *converges to the solution* u *of Problem* 9.16, *i.e.,*

$$u_\rho \to u \quad \text{in } C([0,T], V) \quad \text{as } \rho \to 0. \tag{9.74}$$

Proof. We use Theorem 5.11. To this end, we note that the bilinear form a is symmetric, and it follows from (9.6) that it is continuous and V-elliptic, i.e., it satisfies condition (5.3). Next, (9.62)–(9.64) show that conditions (5.5)–(5.7) are satisfied, the last one with $\delta_0 = 0$. Recall that, as seen in the proof of Theorem 9.3, j is a continuous seminorm on V and satisfies (9.15); therefore, we conclude that j satisfies condition (5.63). Finally, it follows from Lemma 9.7 that conditions (5.64) and (3.17) hold, too. Thus, Theorem 9.19 is a direct consequence of Theorem 5.11, which completes the proof. □

We conclude by Theorem 9.19 that problems 9.15 and 9.17 have a unique weak solution. Also, the weak solution of the quasistatic contact problem with Tresca's friction law can be approached by the weak solution of the quasistatic contact problem with regularized friction law for a sufficiently small parameter ρ.

Finally, note that if we replace condition (9.60) with the condition

$$f_0 \in W^{1,\infty}(0,T;L^2(\Omega)), \quad f_2 \in W^{1,\infty}(0,T;L^2(\Gamma_2)), \tag{9.75}$$

then the solution u of Problem 9.15 has the regularity $u \in W^{1,\infty}(0,T;V)$. Indeed, note that (9.75) implies that the element f given by (9.61) satisfies $f \in W^{1,\infty}(0,T;V)$ and, since j is a continuous seminorm, the existence of a unique solution $u \in W^{1,\infty}(0,T;V)$ to Problem 9.15 follows now from Corollary 5.20. And we also note that in this case, the solution depends Lipschitz continuously on the data f_0, f_2, and u_0 since, using again Corollary 5.20, it follows that the mapping $(f, u_0) \mapsto u$ is Lipschitz continuous from $W^{1,\infty}(0,T;V) \times V$ to $C([0,T];V)$.

9.4 A Quasistatic Slip-dependent Frictional Problem

In this section, we study the quasistatic version of the slip-dependent frictional contact problem presented in Section 9.2. We show that the problem leads to an evolutionary quasivariational inequality for the displacement field and therefore we apply Theorem 5.15 to prove its weak solvability.

The classical formulation of the problem is the following.

Problem 9.20. *Find a displacement field $u : \Omega \times [0,T] \to \mathbb{R}$ such that*

$$\operatorname{div}(\mu \nabla u) + f_0 = 0 \quad \text{in } \Omega \times (0,T), \tag{9.76}$$

$$u = 0 \quad \text{on } \Gamma_1 \times (0,T), \tag{9.77}$$

$$\mu\, \partial_\nu u = f_2 \quad \text{on } \Gamma_2 \times (0,T), \tag{9.78}$$

$$|\mu\, \partial_\nu u| \le g(|u|), \quad \mu\, \partial_\nu u = -g(|u|)\frac{\dot u}{|\dot u|} \quad \text{if } \dot u \neq 0 \quad \text{on } \Gamma_3 \times (0,T), \tag{9.79}$$

$$u(0) = u_0 \quad \text{in } \Omega. \tag{9.80}$$

In the study of this problem, we assume that the Lamé coefficient μ satisfies condition (9.6) and the friction bound g satisfies condition (8.69); moreover, the body forces and surface tractions satisfy

$$f_0 \in W^{1,\infty}(0,T;L^2(\Omega)), \quad f_2 \in W^{1,\infty}(0,T;L^2(\Gamma_2)). \tag{9.81}$$

We use notation (9.8) for the bilinear form a, (9.51) for the functional j, and (9.61) for the function f. It follows from (9.81) that

$$f \in W^{1,\infty}(0,T;V). \tag{9.82}$$

Finally, we assume that the initial displacement u_0 satisfies

$$u_0 \in V, \tag{9.83}$$

$$a(u_0,\, v) + j(u_0,\, v) \geq (f(0),\, v)_V \quad \forall\, v \in V, \tag{9.84}$$

and we recall that (9.84) represents a compatibility condition on the initial data that is necessary to guarantee that initially the elastic body is in equilibrium.

Proceeding as in the case of the static problem 9.1, we obtain the following variational formulation of the contact problem (9.76)–(9.80).

Problem 9.21. *Find a displacement field $u : [0,T] \to V$ such that*

$$\begin{aligned} a(u(t), v - \dot{u}(t)) + j(u(t), v) - j(u(t), \dot{u}(t)) \\ \geq (f(t), v - \dot{u}(t))_V \quad \forall\, v \in V, \quad \text{a.e. } t \in (0,T), \end{aligned} \tag{9.85}$$

$$u(0) = u_0. \tag{9.86}$$

In the study of Problem 9.21, we have the following existence result.

Theorem 9.22. *Assume that conditions* (9.6), (9.81), (9.83), (9.84), *and* (8.69) *hold. Then there exists L_0 that depends only on Ω, Γ_1, Γ_3, and μ such that there exists at least one solution u for Problem* 9.21 *that satisfies $u \in W^{1,\infty}(0,T;V)$, if $L_g < L_0$.*

Proof. The bilinear form a is symmetric, and recall that it satisfies the inequalities (8.51) and (8.52), i.e., is continuous and V-elliptic. It also follows from (9.51) that for all $\eta \in V$, $j(\eta, \cdot)$ is a continuous seminorm on V and, as it was shown in the proof of Theorem 9.14, it satisfies condition (5.101).

Let u, $v \in V$; by using (9.51) and (8.69) we find that

$$\begin{aligned} j(u, v-u) - j(v, v-u) &= \int_{\Gamma_3} (g(|u|) - g(|v|))\,|v-u|\, da \\ &\leq \int_{\Gamma_3} L_g |u-v|^2\, da \leq L_g \|u-v\|^2_{L^2(\Gamma_3)} \end{aligned}$$

and, therefore, by (8.33) we obtain

$$j(u, v-u) - j(v, v-u) \leq c_0^2\, L_g\, \|u-v\|_V^2. \tag{9.87}$$

Let L_0 be defined by (9.54). Clearly, L_0 depends only on Ω, Γ_1, Γ_3, and μ. It follows from (9.87) that j satisfies the inequality in (5.102) with $\alpha = c_0^2\, L_g$, and it follows from (8.52) that a satisfies condition (5.3)(b) with $m = \mu^*$. Assume that $L_g < L_0$; then, we obtain

$$c_0^2\, L_g < \mu^* \tag{9.88}$$

and we conclude by (9.87) that j satisfies condition (5.102).

Let now η, $u \in V$; it follows from (9.51), (8.69), and (8.33) that

$$\begin{aligned} |j(\eta,u)| &\le \int_{\Gamma_3} (L_g|\eta| + g(0))|u|\, da \\ &\le \big(L_g\|\eta\|_{L^2(\Gamma_3)} + \|g(0)\|_{L^2(\Gamma_3)}\big)\|u\|_{L^2(\Gamma_3)} \\ &\le c_0\big(c_0 L_g\|\eta\|_V + \|g(0)\|_{L^2(\Gamma_3)}\big)\|u\|_V. \end{aligned}$$

We use the inequality

$$ab \le \frac{c}{2}\, a^2 + \frac{1}{2c}\, b^2 \quad \forall\, a,\, b,\, c > 0$$

with $a = c_0\big(c_0 L_g\|\eta\|_V + \|g(0)\|_{L^2(\Gamma_3)}\big)$, $b = \|u\|_V$, and $c > 0$ to find that

$$j(\eta,u) \le \frac{c}{2}\, c_0^2\big(c_0 L_g\|\eta\|_V + \|g(0)\|_{L^2(\Gamma_3)}\big)^2 + \frac{1}{2c}\, \|u\|_V^2. \tag{9.89}$$

We now choose c such that

$$c > \frac{1}{2(\mu^* - c_0^2 L_g)} \tag{9.90}$$

and we recall that this choice is allowed by (9.88). Inequality (9.89) shows that (5.103) holds with a_1, $a_2 : V \to \mathbb{R}$ given by

$$a_1(\eta) = \frac{1}{2c}, \quad a_2(\eta) = \frac{c}{2}\, c_0^2\, \big(c_0 L_g\|\eta\|_V + \|g(0)\|_{L^2(\Gamma_3)}\big)^2,$$

as by (9.90) we have $a_1(0_V) = \frac{1}{2c} < \mu^* - c_0^2 L_g$.

Let now $\{\eta_n\} \subset V$ be such that $\eta_n \rightharpoonup \eta \in V$ and let $\{u_n\} \subset V$ be a bounded sequence, that is, there exists $c > 0$ such that

$$\|u_n\|_V \le c \quad \forall\, n \in \mathbb{N}. \tag{9.91}$$

We use (9.51), (8.69), and (8.33) to find that

$$|j(\eta_n, u_n) - j(\eta, u_n)| \le c_0 L_g \|\eta_n - \eta\|_{L^2(\Gamma_3)} \|u_n\|_V \quad \forall\, n \in \mathbb{N}. \tag{9.92}$$

Using the compactness property of the trace map, it follows that

$$\|\eta_n - \eta\|_{L^2(\Gamma_3)} \to 0. \tag{9.93}$$

We combine now (9.91)–(9.93) to see that j satisfies condition (5.104).

We conclude that a and j satisfy conditions (5.3), (5.100)–(5.104) of Theorem 5.15 and we recall that V is a separable Hilbert space. Therefore, since (9.82)–(9.84) hold, it follows from Theorem 5.15(1) that there exists at least a solution $u \in W^{1,\infty}(0,T;V)$ to Problem 9.21. □

Theorem 9.22 shows the solvability of the Problem 9.21, if the Lipschitz constant L_g is small enough. Using (9.54), it follows that the critical value $L_0 = \frac{\mu}{c_0^2}$ (where c_0 is defined on page 153) depends only on Ω, Γ_1, Γ_3, and μ, and does not depend on the other data of the problem. Note that the critical value L_0 in Theorems 9.13 and 9.22 is the same. However, its significance in the theorems above is different. Indeed, the existence of a weak solution for the static antiplane slip-dependent frictional contact problem 9.12 is guaranteed by Theorem 9.14 and, as it results from Theorem 9.13, the smallness assumption $L_g < L_0$ represents a sufficient condition to derive its *uniqueness.* The situation is different concerning the quasistatic antiplane slip-dependent frictional contact problem 9.21 since $L_g < L_0$ represents a sufficient condition to derive the *existence* of the weak solution, see Theorem 9.22. The question if this smallness assumption represents an intrinsic feature of the above antiplane contact problems or represents only a limitation of our mathematical tools is left open and, clearly, it needs further investigation in the future.

Chapter 10
Viscoelastic Problems with Short Memory

In this chapter, we study quasistatic models for antiplane frictional problems involving viscoelastic materials with short memory. We model the friction with versions of Coulomb's law, including Tresca's law and its regularization, slip-dependent and total slip-dependent laws. For all the models, we prove that the displacement field satisfies an evolutionary variational inequality with viscosity. Then we apply the results in Chapter 4 to obtain existence, uniqueness, regularity, and convergence results. In this chapter, we use again the space V (page 152), together with its inner product $(\cdot,\cdot)_V$ and the associated norm $\|\cdot\|_V$. Moreover, $[0,T]$ represents the time interval of interest, $T>0$. Finally, we note that everywhere in this chapter, the use of the abstract results presented in Part II of this book is made in the case $X=V$, $(\cdot,\cdot)_X=(\cdot,\cdot)_V$, without explicit specification.

10.1 Problems with Tresca and Regularized Friction

For the first antiplane frictional contact problem we consider in this section, the friction is modeled with Tresca's law and, therefore, the classical model of the process is the following.

Problem 10.1. *Find a displacement field $u:\Omega\times[0,T]\to\mathbb{R}$ such that*

$$\operatorname{div}(\theta\,\nabla\dot u+\mu\,\nabla u)+f_0=0 \quad \text{in } \Omega\times(0,T), \tag{10.1}$$

$$u=0 \quad \text{on } \Gamma_1\times(0,T), \tag{10.2}$$

$$\theta\,\partial_\nu\dot u+\mu\,\partial_\nu u=f_2 \quad \text{on } \Gamma_2\times(0,T), \tag{10.3}$$

$$\left.\begin{array}{l} |\theta\,\partial_\nu\dot u+\mu\,\partial_\nu u|\le g, \\ \theta\,\partial_\nu\dot u+\mu\,\partial_\nu u=-g\,\frac{\dot u}{|\dot u|} \quad \text{if } \dot u\neq 0 \end{array}\right\} \quad \text{on } \Gamma_3\times(0,T), \tag{10.4}$$

$$u(0)=u_0 \quad \text{in } \Omega. \tag{10.5}$$

Note that (10.1) represents the equation of equilibrium (8.11), (10.2) represents the displacement boundary condition (8.16), (10.3) represents the surface traction boundary condition (8.24), (10.4) represents the Tresca friction law (8.61), and, finally, (10.5) represents the initial condition. Note also that with respect to the quasistatic antiplane problem with Tresca's friction law treated in Section 9.3, the equilibrium equation (10.1) as well as the boundary conditions (10.3) and (10.4) contain now an additional term involving the viscosity coefficient θ; this term arises from the fact that here we use the viscoelastic constitutive law with short memory instead of the elastic constitutive law used everywhere in Chapter 9.

In the study of Problem 10.1, we assume that the elasticity and the viscosity coefficients satisfy

$$\mu \in L^\infty(\Omega), \tag{10.6}$$

$$\theta \in L^\infty(\Omega), \text{ and there exists } \theta^* > 0$$
$$\text{such that } \theta(\boldsymbol{x}) \geq \theta^* \text{ a.e. } \boldsymbol{x} \in \Omega. \tag{10.7}$$

The body forces and surface tractions densities satisfy

$$f_0 \in C([0,T]; L^2(\Omega)), \quad f_2 \in C([0,T]; L^2(\Gamma_2)), \tag{10.8}$$

the friction bound g satisfies

$$g \in L^2(\Gamma_3) \text{ and } g(\boldsymbol{x}) \geq 0 \quad \text{a.e. } \boldsymbol{x} \in \Gamma_3, \tag{10.9}$$

and, finally, the initial displacement is such that

$$u_0 \in V. \tag{10.10}$$

We define the bilinear forms $a : V \times V \to \mathbb{R}$, $b : V \times V \to \mathbb{R}$ and the functional $j : V \to \mathbb{R}$ by the equalities

$$a(u,v) = \int_\Omega \mu \nabla u \cdot \nabla v \, dx \quad \forall\, u,\, v \in V, \tag{10.11}$$

$$b(u,v) = \int_\Omega \theta \nabla u \cdot \nabla v \, dx \quad \forall\, u,\, v \in V, \tag{10.12}$$

$$j(v) = \int_{\Gamma_3} g|v| \, da \quad \forall\, v \in V, \tag{10.13}$$

and we note that by assumptions (10.6)–(10.7) and (10.9), the integrals in (10.11)–(10.13) are well defined. We also use (10.8) and the Riesz representation theorem (page 11) to define the function $f : [0,T] \to V$ by the equality

$$(f(t), v)_V = \int_\Omega f_0(t)\, v \, dx + \int_{\Gamma_2} f_2(t)\, v \, da \quad \forall\, v \in V,\ t \in [0,T]. \tag{10.14}$$

It follows from (10.8) that the integrals in (10.14) are well defined and, moreover,

$$f \in C([0,T];V). \tag{10.15}$$

With these notation, a standard calculation based on Green's formula (8.35) as well as on the arguments presented in Section 9.1 leads to the following variational formulation of the frictional problem (10.1)–(10.5).

Problem 10.2. *Find a displacement field $u : [0,T] \to V$ such that*

$$\begin{aligned} &a(u(t), v - \dot{u}(t)) + b(\dot{u}(t), v - \dot{u}(t)) + j(v) - j(\dot{u}(t)) \\ &\quad \geq (f(t), v - \dot{u}(t))_V \quad \forall\, v \in V,\ t \in [0,T], \end{aligned} \tag{10.16}$$

$$u(0) = u_0. \tag{10.17}$$

In the study of Problem 10.2, we have the following existence and uniqueness result.

Theorem 10.3. *Assume* (10.6)–(10.10). *Then, Problem* 10.2 *has a unique solution $u \in C^1([0,T];V)$.*

Proof. It follows from condition (10.6) that the bilinear form a satisfies

$$|a(u,v)| \leq \|\mu\|_{L^\infty(\Omega)} \|u\|_V \|v\|_V \quad \forall\, u,\, v \in V. \tag{10.18}$$

Also, the bilinear form b is symmetric and, by assumption (10.7) on the viscosity coefficient θ, we deduce that

$$|b(u,v)| \leq \|\theta\|_{L^\infty(\Omega)} \|u\|_V \|v\|_V \quad \forall\, u,\, v \in V, \tag{10.19}$$

$$b(v,v) \geq \theta^* \|v\|_V^2 \quad \forall\, v \in V. \tag{10.20}$$

Note also that (10.9) implies that functional j is a continuous seminorm on V and, therefore, is convex and lower semicontinuous. These properties combined with the regularity (10.15) and (10.10) allow us to apply Theorem 4.1 to problem (10.16)–(10.17) in order to conclude the proof. □

We now establish a regularity result.

Theorem 10.4. *Under the conditions stated in Theorem* 10.3, *if the body forces and surface tractions densities have the regularity*

$$f_0 \in W^{1,p}(0,T;L^2(\Omega)), \quad f_2 \in W^{1,p}(0,T;L^2(\Gamma_2)) \tag{10.21}$$

for some $p \in [1,\infty]$, then $u \in W^{2,p}(0,T;V)$.

Proof. We use the definition (10.14) to see that assumption (10.21) implies the regularity $f \in W^{1,p}(0,T;V)$. Theorem 10.4 is now a consequence of Theorem 4.5 in Section 4.1. □

We conclude by Theorem 10.3 that the mechanical problem 10.1 has a unique weak solution. The regularity of the solution is higher than the regularity of the data, since if f_0 and f_2 are continuous, then the solution u is continuously differentiable, and if f_0 and f_2 are $W^{1,p}$ functions, then the solution u is a $W^{2,p}$ function, as derived in Theorem 10.4. This feature arises from the structure of the problem that involves a new bilinear form, b, in which the derivative $\dot{u}$ of the solution plays the main role. Note also that, in comparison with Theorem 9.19, Theorems 10.3 and 10.4 provide a better regularity result, under a less restrictive assumption on the Lamé coefficient μ and on the initial displacement u_0 (compare (9.6) with (10.6) and (9.63)–(9.64) with (10.10), respectively). We conclude that adding viscosity has a *smoothing effect* on the solution of antiplane frictional contact problems.

We now consider the viscoelastic problem with regularized friction. The classical formulation of the problem is as follows.

Problem 10.5. *Find a displacement field* $u_\rho : \Omega \times [0,T] \to \mathbb{R}$ *such that*

$$\operatorname{div}(\theta\,\nabla\dot{u}_\rho + \mu\,\nabla u_\rho) + f_0 = 0 \qquad \text{in } \Omega \times (0,T), \tag{10.22}$$

$$u_\rho = 0 \qquad \text{on } \Gamma_1 \times (0,T), \tag{10.23}$$

$$\theta\,\partial_\nu\dot{u}_\rho + \mu\,\partial_\nu u_\rho = f_2 \qquad \text{on } \Gamma_2 \times (0,T), \tag{10.24}$$

$$\theta\,\partial_\nu\dot{u}_\rho + \mu\,\partial_\nu u_\rho = -g\,\frac{\dot{u}_\rho}{\sqrt{\dot{u}_\rho^2 + \rho^2}} \qquad \text{on } \Gamma_3 \times (0,T), \tag{10.25}$$

$$u_\rho(0) = u_0 \qquad \text{in } \Omega. \tag{10.26}$$

This is the viscoelastic version of the quasistatic elastic problem 9.17, studied in Section 9.3. In the study of this problem, we assume that (10.6)–(10.10) hold and, moreover, $\rho > 0$. We use notation (10.11), (10.12), and (10.14) above and we introduce the frictional functional $j_\rho : V \to \mathbb{R}$ given by

$$j_\rho(v) = \int_{\Gamma_3} g\Big(\sqrt{v^2 + \rho^2} - \rho\Big)\,da \quad \forall\, v \in V. \tag{10.27}$$

Then, the variational formulation of the mechanical problem 10.5 is the following.

Problem 10.6. *Find a displacement field* $u_\rho : [0,T] \to V$ *such that*

$$a(u_\rho(t), v - \dot{u}_\rho(t)) + b(\dot{u}_\rho(t), v - \dot{u}_\rho(t)) + j_\rho(v) - j_\rho(\dot{u}_\rho(t)) \\ \geq (f(t), v - \dot{u}_\rho(t))_V \quad \forall\, v \in V,\ t \in [0,T], \tag{10.28}$$

$$u(0) = u_0. \tag{10.29}$$

It follows from Lemma 9.7 and Corollary 1.37 (page 14) that j_ρ is a convex l.s.c. functional on V and j_ρ and j satisfy condition (3.17). Therefore, using again Theorems 4.1, 4.5, and 4.6, we obtain the following result.

Theorem 10.7. *Assume* (10.6)–(10.10). *Then, for every* $\rho > 0$, *Problem* 10.6 *has a unique solution* $u_\rho \in C^1([0,T];V)$, *and* $u_\rho \in W^{2,p}(0,T;V)$ *if* (10.21) *holds. Moreover, the solution converges to the solution* u *of Problem* 10.2, *i.e.,*

$$\|u_\rho - u\|_{C^1([0,T];V)} \to 0 \quad \text{as } \rho \to 0. \tag{10.30}$$

We conclude by Theorem 10.7 that the antiplane frictional problem (10.22)–(10.26) has a unique weak solution that converges to the weak solution of Problem 10.1 as $\rho \to 0$. In addition to the interest from the asymptotic point of view, this result is important from the mechanical point of view since it shows that, in the study of antiplane frictional contact problems for viscoelastic materials with short memory, the solution of the problem with the Tresca friction law can be approached as closely as one wishes by the solution of the problem with regularized friction law with a sufficiently small parameter ρ. Moreover, since the functional j_ρ defined in (10.27) is Gâteaux differentiable, by using the arguments presented on page 67 it follows that solving the variational inequality (10.28)–(10.29) leads to the solution of a Cauchy problem for nonlinear evolution equations on the space V, which is convenient from the numerical point of view.

10.2 Approach to Elasticity

In this section, we investigate the behavior of the weak solution of the antiplane frictional problems in Section 10.1 as the viscosity converges to zero. Our results below are valid both for the Tresca friction law and its regularization; however, to avoid repetitions, we restrict ourselves to the case of the Tresca friction law. In order to outline the dependence on the viscosity coefficient θ, we use in this section the notation b_θ for the bilinear form (10.12), i.e.,

$$b_\theta(u,v) = \int_\Omega \theta\, \nabla u \cdot \nabla v \, dx, \quad \forall\, u,\, v \in V,$$

and reformulate Problem 10.2 as follows.

Problem 10.8. *Find a displacement field* $u_\theta : [0,T] \to V$ *such that*

$$a(u_\theta(t), v - \dot u_\theta(t)) + b_\theta(\dot u_\theta(t),\, v - \dot u_\theta(t)) + j(v) - j(\dot u_\theta(t)) \\ \geq (f(t), v - \dot u_\theta(t))_V \quad \forall\, v \in V, \text{ a.e. } t \in (0,T), \tag{10.31}$$

$$u_\theta(0) = u_0. \tag{10.32}$$

We also consider the inviscid problem associated to (10.31)–(10.32), i.e., the problem obtained for $\theta = 0$. Note that this problem is the elastic frictional problem 9.16 studied in Section 9.3 and is formulated as follows.

Problem 10.9. *Find a displacement field* $u : [0,T] \to V$ *such that*

$$a(u(t), v - \dot{u}(t)) + j(v) - j(\dot{u}(t)) \\ \geq (f(t), v - \dot{u}(t))_V \quad \forall v \in V, \text{ a.e. } t \in (0,T), \tag{10.33}$$

$$u(0) = u_0. \tag{10.34}$$

Assume in what follows that (9.6), (10.7), (10.9), (9.60), (9.63), and (9.64) hold. Then, it follows from Theorem 10.4 that Problem (10.31)–(10.32) has a unique solution $u_\theta \in W^{2,2}(0,T;V)$. It also follows from Theorem 9.19 that problem (10.33)–(10.34) has a unique solution $u \in W^{1,2}(0,T;V)$. Consider now the assumption

$$\frac{1}{\theta^*} \|\theta\|^2_{L^\infty(\Omega)} \to 0. \tag{10.35}$$

We have the following convergence result.

Theorem 10.10. *Under the stated assumptions, if* (10.35) *holds, then the solution* u_θ *of Problem* 10.8 *converges to the solution* u *of Problem* 10.9, *i.e.,*

$$\|u_\theta - u\|_{C([0,T];V)} \to 0. \tag{10.36}$$

Proof. It follows from (10.19) and (10.20) that the bilinear form b_θ satisfies condition (5.88) with $M'(\theta) = \|\theta\|_{L^\infty(\Omega)}$ and $m'(\theta) = \theta^*$. The convergence result (10.36) is now a consequence of Theorem 5.14. □

Consider now the case of homogeneous viscosity, i.e., the case when assumption (10.7) is replaced by the assumption

$$\theta(x) = \theta > 0 \quad \text{a.e. } \boldsymbol{x} \in \Omega,$$

where θ is given. In this case, $\|\theta\|_{L^\infty(\Omega)} = \theta$, $\theta^* = \theta$ and so the convergence (10.35) is equivalent to $\theta \to 0$. Therefore, by Theorem 10.10 we conclude that the weak solution of the antiplane viscoelastic problem with Tresca's friction law may be approached by the weak solution of the antiplane elastic problem with Tresca's friction law, as the viscosity is small enough. From the mechanical point of view, this convergence result shows that the elasticity with friction may be considered as a limit case of viscoelasticity with friction as the viscosity decreases.

10.3 Slip- and Slip Rate–dependent Frictional Problems

In this section, we consider the frictional contact problem 10.1 in the case when the friction bound g depends on the slip $|u|$ or on the slip rate $|\dot{u}|$.

The classical formulation of the problems follows in a straightforward way from (10.1)–(10.5), by replacing g with $g(|u|)$ or $g(|\dot{u}|)$, respectively. Thus, the antiplane viscoelastic contact problem with slip-dependent friction law is formulated as follows.

Problem 10.11. *Find a displacement field* $u : \Omega \times [0,T] \to \mathbb{R}$ *such that*

$$\operatorname{div}(\theta\,\nabla\dot{u} + \mu\,\nabla u) + f_0 = 0 \quad \text{in } \Omega \times (0,T), \tag{10.37}$$

$$u = 0 \quad \text{on } \Gamma_1 \times (0,T), \tag{10.38}$$

$$\theta\,\partial_\nu \dot{u} + \mu\,\partial_\nu u = f_2 \quad \text{on } \Gamma_2 \times (0,T), \tag{10.39}$$

$$\left.\begin{array}{l} |\theta\,\partial_\nu \dot{u} + \mu\,\partial_\nu u| \le g(|u|), \\ \theta\,\partial_\nu \dot{u} + \mu\,\partial_\nu u = -g(|u|)\,\frac{\dot{u}}{|\dot{u}|} \quad \text{if } \dot{u} \ne 0 \end{array}\right\} \quad \text{on } \Gamma_3 \times (0,T), \tag{10.40}$$

$$u(0) = u_0 \quad \text{in } \Omega. \tag{10.41}$$

Also, the classical formulation of the antiplane viscoelastic contact problem with slip rate–dependent friction law is as follows.

Problem 10.12. *Find a displacement field* $u : \Omega \times [0,T] \to \mathbb{R}$ *such that*

$$\operatorname{div}(\theta\,\nabla\dot{u} + \mu\,\nabla u) + f_0 = 0 \quad \text{in } \Omega \times (0,T), \tag{10.42}$$

$$u = 0 \quad \text{on } \Gamma_1 \times (0,T), \tag{10.43}$$

$$\theta\,\partial_\nu \dot{u} + \mu\,\partial_\nu u = f_2 \quad \text{on } \Gamma_2 \times (0,T), \tag{10.44}$$

$$\left.\begin{array}{l} |\theta\,\partial_\nu \dot{u} + \mu\,\partial_\nu u| \le g(|\dot{u}|), \\ \theta\,\partial_\nu \dot{u} + \mu\,\partial_\nu u = -g(|\dot{u}|)\,\frac{\dot{u}}{|\dot{u}|} \quad \text{if } \dot{u} \ne 0 \end{array}\right\} \quad \text{on } \Gamma_3 \times (0,T), \tag{10.45}$$

$$u(0) = u_0 \quad \text{in } \Omega. \tag{10.46}$$

The variational formulation of Problem 10.11 can be obtained by using the arguments presented in Section 9.1 and is the following.

Problem 10.13. *Find a displacement field* $u : [0,T] \to V$ *such that*

$$\begin{aligned} &a(u(t), v - \dot{u}(t)) + b(\dot{u}(t), v - \dot{u}(t)) + j(u(t), v) - j(u(t), \dot{u}(t)) \\ &\quad \ge (f(t), v - \dot{u}(t))_V \quad \forall\, v \in V,\ t \in [0,T], \end{aligned} \tag{10.47}$$

$$u(0) = u_0. \tag{10.48}$$

Here and below in this section, we use notation (10.11) and (10.12) for the bilinear forms a and b as well as notation (10.14) for the function f. Also, the functional $j : V \times V \to \mathbb{R}$ is given by

$$j(u, v) = \int_{\Gamma_3} g(|u|)\, |v|\, da \quad \forall\, u,\, v \in V, \tag{10.49}$$

where the friction bound g is assumed to satisfy (8.69).

The variational formulation of Problem 10.12 is the following.

Problem 10.14. *Find a displacement field $u : [0, T] \to V$ such that*

$$\begin{aligned} &a(u(t), v - \dot{u}(t)) + b(\dot{u}(t), v - \dot{u}(t)) + j(\dot{u}(t), v) - j(\dot{u}(t), \dot{u}(t)) \\ &\quad \geq (f(t), v - \dot{u}(t))_V \quad \forall\, v \in V,\ t \in [0, T], \end{aligned} \tag{10.50}$$

$$u(0) = u_0. \tag{10.51}$$

The well-posedness of the variational problems 10.13 and 10.14 is given by the following existence and uniqueness result.

Theorem 10.15. *Assume* (10.6)–(10.8), (10.10), *and* (8.69). *Then:*

(1) There exists a unique solution $u \in C^1([0, T]; V)$ to problem (10.47)–(10.48).

(2) There exists L_0, which depends on Ω, Γ_1, Γ_3, and θ, such that there exists a unique solution $u \in C^1([0, T]; V)$ to problem (10.50)–(10.51), *if $L_g < L_0$.*

Proof. The bilinear form a, the function f, and the initial data u_0 satisfy conditions (4.3), (4.6), and (4.7), respectively. In addition, for all $\eta \in V$, the functional $j(\eta, \cdot) : V \to \mathbb{R}$ is a continuous seminorm on V and therefore it satisfies condition (3.30)(a). Recall also that j satisfies inequality (9.53) where c_0 is given in (8.33), which shows that condition (3.30)(b) holds with $\alpha = c_0^2\, L_g$. Moreover, it follows from (10.19) and (10.20) that the bilinear form b satisfies condition (4.4) with $m' = \theta^*$. Choose $L_0 = \frac{\theta^*}{c_0^2}$, which depends on Ω, Γ_1, Γ_3, and θ. Then, if $L_g < L_0$ we have $m' > \alpha$, and therefore Theorem 10.15 is a direct consequence of Theorem 4.7. □

It follows from Theorem 10.15 that the frictional viscoelastic problem with slip-dependent friction law has a unique weak solution and, under a smallness assumption, the frictional viscoelastic problem with slip rate–dependent friction law has a unique weak solution, too. Note also that the critical value L_0 depends only on the viscosity coefficient θ and on the geometry of the problem, but does not depend on the Lamé coefficient μ, the external forces, nor on the initial displacement.

Finally, we remark that if the body forces and surface tractions densities satisfy (10.21) for some $p \in [1, \infty]$, then the solution of problems 10.13 and 10.14 satisfies $u \in W^{2,p}(0, T; V)$. Indeed, this regularity is a direct consequence of Theorem 4.8.

10.4 Total Slip- and Total Slip Rate–dependent Frictional Problems

For the problems we study in this section, the friction bound g is assumed to depend on the total slip or on the total slip rate and therefore g satisfies (8.66) or (8.67), respectively. For this reason, in what follows, for every function $v \in C([0,T];V)$ we denote by $Sv(t)$ the element in $L^2(\Gamma_3)$ given by

$$Sv(t) = \int_0^t |v(s)|\, ds \quad \forall\, t \in [0,T], \tag{10.52}$$

and we note that here and below we do not indicate the dependence of functions on $\boldsymbol{x} \in \Omega \cup \Gamma_3$. Also, note that in (10.52) we write $|v(s)|$ for the absolue value of the trace of the function $v(s)$ on Γ_3, which is an element of $L^2(\Gamma_3)$, and the integral above is understood in the space $L^2(\Gamma_3)$. As a result $Sv(t) \in L^2(\Gamma_3)$, and S is an operator that maps every function $v \in C([0,T];V)$ to the function $Sv \in C([0,T], L^2(\Gamma_3))$, i.e., $S : C([0,T];V) \to C([0,T], L^2(\Gamma_3))$.

We use now (10.52) in (8.66) and (8.67) to see that the friction bound function is of the form $g(Su(t))$ in the case of the problem with total slip-dependent friction and of the form $g(S\dot{u}(t))$ in the case of the problem with total slip rate–dependent friction. Therefore, the classical formulation of the problems follows directly from (10.1)–(10.5) by replacing g with $g(Su(t))$ or $g(S\dot{u}(t))$, respectively. Thus, the antiplane viscoelastic contact problem with total slip-dependent friction law is formulated as follows.

Problem 10.16. *Find a displacement field $u : \Omega \times [0,T] \to \mathbb{R}$ such that*

$$\operatorname{div}\,(\theta\,\nabla \dot{u}(t) + \mu\,\nabla u(t)) + f_0(t) = 0 \quad \text{in } \Omega, \tag{10.53}$$

$$u(t) = 0 \quad \text{on } \Gamma_1, \tag{10.54}$$

$$\theta\,\partial_\nu \dot{u}(t) + \mu\,\partial_\nu u(t) = f_2(t) \quad \text{on } \Gamma_2, \tag{10.55}$$

$$\left.\begin{aligned} &|\theta\,\partial_\nu \dot{u}(t) + \mu\,\partial_\nu u(t)| \le g(Su(t)),\\ &\theta\,\partial_\nu \dot{u}(t) + \mu\,\partial_\nu u(t) = -g(Su(t))\,\frac{\dot{u}(t)}{|\dot{u}(t)|} \quad \text{if } \dot{u}(t) \neq 0 \end{aligned}\right\} \quad \text{on } \Gamma_3, \tag{10.56}$$

for all $t \in [0,T]$,

$$u(0) = u_0 \quad \text{in } \Omega. \tag{10.57}$$

The classical formulation of the antiplane viscoelastic contact problem with total slip rate–dependent friction law is as follows.

Problem 10.17. *Find a displacement field* $u : \Omega \times [0, T] \to \mathbb{R}$ *such that*

$$\mathrm{div}\,(\theta\,\nabla\dot{u}(t) + \mu\,\nabla u(t)) + f_0(t) = 0 \quad \text{in } \Omega, \tag{10.58}$$

$$u(t) = 0 \quad \text{on } \Gamma_1, \tag{10.59}$$

$$\theta\,\partial_\nu \dot{u}(t) + \mu\,\partial_\nu u(t) = f_2(t) \quad \text{on } \Gamma_2, \tag{10.60}$$

$$\left.\begin{array}{l} |\theta\,\partial_\nu \dot{u}(t) + \mu\,\partial_\nu u(t)| \le g(S\dot{u}(t)), \\ \theta\,\partial_\nu \dot{u}(t) + \mu\,\partial_\nu u(t) = -g(S\dot{u}(t))\,\frac{\dot{u}(t)}{|\dot{u}(t)|} \quad \text{if } \dot{u}(t) \ne 0 \end{array}\right\} \quad \text{on } \Gamma_3, \tag{10.61}$$

for all $t \in [0, T]$,

$$u(0) = u_0 \quad \text{in } \Omega. \tag{10.62}$$

The variational formulation of Problem 10.16 can be obtained by using the arguments presented in Section 9.1 and is the following.

Problem 10.18. *Find a displacement field* $u : [0, T] \to V$ *such that*

$$\begin{aligned} &a(u(t), v - \dot{u}(t)) + b(\dot{u}(t), v - \dot{u}(t)) + j(Su(t), v) - j(Su(t), \dot{u}(t)) \\ &\quad \ge (f(t), v - \dot{u}(t))_V \quad \forall\, v \in V,\ t \in [0, T], \end{aligned} \tag{10.63}$$

$$u(0) = u_0. \tag{10.64}$$

Here and everywhere in this section, we use notation (10.11), (10.12), (10.14), and (10.52). Also, $j : L^2(\Gamma_3) \times V \to \mathbb{R}$ is the functional given by

$$j(\eta, v) = \int_{\Gamma_3} g(\eta)\,|v|\,da \quad \forall\, \eta \in L^2(\Gamma_3),\ v \in V. \tag{10.65}$$

The integral above exists under condition (8.69) on g, which we assume in this section. We need to define j as above since at each time moment $t \in [0, T]$, the total slip $Su(t)$ belongs to the space $L^2(\Gamma_3)$. Considering the functional (10.65) instead of the functional (10.49) represents the first difference with respect to the slip-dependent frictional problem studied in Section 10.3; the second difference consists in the fact that now the problem is history dependent since, unlike the functional j in Problem 10.13 that depends on the current slip $|u(t)|$, the functional j in Problem 10.18 depends on the history of the slip, $Su(t)$.

Also, with the notation above, the variational formulation of Problem 10.17 is as follows.

Problem 10.19. *Find a displacement field* $u : [0, T] \to V$ *such that*

$$\begin{aligned} &a(u(t), v - \dot{u}(t)) + b(\dot{u}(t), v - \dot{u}(t)) + j(S\dot{u}(t), v) - j(S\dot{u}(t), \dot{u}(t)) \\ &\quad \ge (f(t), v - \dot{u}(t))_V \quad \forall\, v \in V,\ t \in [0, T], \end{aligned} \tag{10.66}$$

$$u(0) = u_0. \tag{10.67}$$

Again, note that considering the functional (10.65) instead of the functional (10.49) represents the first difference with respect to the slip rate–dependent frictional problem studied in Section 10.3; the second difference consists in the fact that now the problem is history dependent, since the functional j in Problem 10.19 depends on the history of the slip rate, $S\dot{u}(t)$.

The well-posedness of the variational problems 10.18 and 10.19 is given by the following existence and uniqueness results.

Theorem 10.20. *Assume* (10.6)–(10.8), (10.10), *and* (8.69) *then:*

(1) There exists a unique solution $u \in C^1([0,T];V)$ *to problem* (10.63)–(10.64).

(2) There exists a unique solution $u \in C^1([0,T];V)$ *to problem* (10.66)–(10.67).

Proof. Conditions (10.6) and (10.7) imply that the bilinear forms a and b defined by (10.11) and (10.12) satisfy assumptions (4.3) and (4.4), respectively; condition (8.69) on the friction bound g implies that the functional j given by (10.65) satisfies assumption (3.54) with $X = V$ and $Y = L^2(\Gamma_3)$; also, it follows from (10.52) that

$$\|Sv_1(t) - Sv_2(t)\|_{L^2(\Gamma_3)} \\ \leq \int_0^t \|v_1(s) - v_2(s)\|_{L^2(\Gamma_3)} ds \quad \forall v_1,\, v_2 \in C([0,T];V),\ t \in [0,T]$$

and, using (8.33), we deduce that $S : C([0,T];V) \to C([0,T];L^2(\Gamma_3))$ satisfies condition (4.40) with $X = V$ and $Y = L^2(\Gamma_3)$; finally, recall the regularity (10.15) and (10.10), which shows that (4.6) and (4.7) also hold, respectively. Theorem 10.20 is now a consequence of Theorem 4.9 in Section 4.4. □

We conclude from Theorem 10.20 that the frictional viscoelastic problem with total slip-dependent friction law and the frictional viscoelastic problem with total slip rate–dependent friction law have, each one, a unique weak solution that satisfies problems 10.18 and 10.19, respectively.

When compared with the result of Theorem 10.15(2), in Theorem 10.20(2) we do not need any smallness assumption on the friction bound g. The reason lies in the structure of the corresponding problems: in Problem 10.14, the functional j depends on the current value of the derivative of the solution, $\dot{u}(t)$, whereas in Problem 10.19, the functional j depends on the solution via an integral term, the total slip rate $S\dot{u}(t)$.

Chapter 11
Viscoelastic Problems with Long Memory

For the problems in this chapter, we assume that the behavior of the materials is described by a viscoelastic constitutive law with long memory. We consider both static and quasistatic antiplane problems in which the friction conditions are either the Tresca law or its regularization. For each model, we derive a variational formulation that is in the form of an elliptic or evolutionary variational inequality with a Volterra integral term for the displacement field. Then, by using the abstract results in Chapter 6, we derive existence, uniqueness, and convergence results for the weak solutions of the corresponding antiplane frictional contact problems. In particular, we study the behavior of the solutions as the relaxation coefficient converges to zero and prove that they converge to the solution of the corresponding purely elastic problems. Again, everywhere in this chapter, we use the space V (page 152), endowed with its inner product $(\cdot,\cdot)_V$ and the associated norm $\|\cdot\|_V$, and we denote by $[0,T]$ the time interval of interest, $T>0$. Also, everywhere in this chapter, the use of the abstract results presented in Part II of this manuscript is made in the case $X=V$, $(\cdot,\cdot)_X=(\cdot,\cdot)_V$, without explicit specification.

11.1 Static Frictional Problems

In this section, we study the viscoelastic version of two of the static problems presented in Section 9.1. For the first problem, the contact is modeled with Tresca's friction law and, therefore, the classical model of the process is the following.

Problem 11.1. *Find a displacement field* $u:\Omega\times[0,T]\to\mathbb{R}$ *such that*

$$\operatorname{div}\Big(\mu\,\nabla u(t)+\int_0^t\theta(t-s)\,\nabla u(s)\,ds\Big)+f_0(t)=0\quad\text{in }\Omega,\tag{11.1}$$

$$u(t)=0\quad\text{on }\Gamma_1,\tag{11.2}$$

$$\mu\partial_\nu u(t) + \int_0^t \theta(t-s)\partial_\nu u(s)\,ds = f_2(t) \quad \text{on } \Gamma_2, \tag{11.3}$$

$$\left.\begin{aligned} &\left|\mu\partial_\nu u(t) + \int_0^t \theta(t-s)\partial_\nu u(s)\,ds\right| \le g, \\ &\mu\partial_\nu u(t) + \int_0^t \theta(t-s)\partial_\nu u(s)\,ds \\ &\quad = -g\frac{u(t)}{|u(t)|} \quad \text{if } u(t) \neq 0 \end{aligned}\right\} \quad \text{on } \Gamma_3, \tag{11.4}$$

for all $t \in [0,T]$.

Note that (11.1) represents the equation of equilibrium (8.15), (11.2) is the displacement boundary condition (8.16), (11.3) represents the traction boundary condition (8.25), and, finally, (11.4) represents the static version of Tresca's friction law, (8.62), in the case of viscoelastic materials with long memory.

In the study of Problem 11.1, we assume that the elasticity and relaxation coefficients satisfy

$$\begin{aligned} &\mu \in L^\infty(\Omega) \text{ and there exists } \mu^* > 0 \\ &\quad \text{such that } \mu(\boldsymbol{x}) \ge \mu^* \quad \text{a.e. } \boldsymbol{x} \in \Omega, \end{aligned} \tag{11.5}$$

$$\theta \in C([0,T]; L^\infty(\Omega)). \tag{11.6}$$

The body forces and surface tractions densities satisfy

$$f_0 \in C([0,T]; L^2(\Omega)), \quad f_2 \in C([0,T]; L^2(\Gamma_2)), \tag{11.7}$$

and, finally, the friction bound g satisfies

$$g \in L^2(\Gamma_3) \quad \text{and} \quad g(\boldsymbol{x}) \ge 0 \quad \text{a.e. } \boldsymbol{x} \in \Gamma_3. \tag{11.8}$$

We define the bilinear form $a : V \times V \to \mathbb{R}$ and the functional $j : V \to \mathbb{R}$ by

$$a(u,v) = \int_\Omega \mu \nabla u \cdot \nabla v\,dx \quad \forall\, u,\, v \in V, \tag{11.9}$$

$$j(v) = \int_{\Gamma_3} g|v|\,da \quad \forall\, v \in V, \tag{11.10}$$

and we note that by assumptions (11.5) and (11.8), the integrals in (11.9) and (11.10) are well defined. We also use (11.6), (11.7), and Riesz's representation theorem (page 11) to define the operator $A : [0,T] \to \mathcal{L}(V)$ and the function $f : [0,T] \to V$ by the equalities

$$(A(t)u, v)_V = \int_\Omega \theta(t)\, \nabla u \cdot \nabla v\,dx \quad \forall\, u,\, v \in V,\ t \in [0,T], \tag{11.11}$$

$$(f(t), v)_V = \int_\Omega f_0(t)\, v\,dx + \int_{\Gamma_2} f_2(t)\, v\,da \quad \forall\, v \in V,\ t \in [0,T]. \tag{11.12}$$

It follows from (11.11), (11.12), and (8.32) that

$$\|A(t_1) - A(t_2)\|_{\mathcal{L}(V)} \le \|\theta(t_1) - \theta(t_2)\|_{L^\infty(\Omega)}, \tag{11.13}$$

$$\|f(t_1) - f(t_2)\|_V \le c\left(\|f_0(t_1) - f_0(t_2)\|_{L^2(\Omega)} + \|f_2(t_1) - f_2(t_2)\|_{L^2(\Gamma_2)}\right) \tag{11.14}$$

for all t_1, $t_2 \in [0,T]$, and, keeping in mind (11.6) and (11.7), we deduce that

$$A \in C([0,T];\mathcal{L}(V)), \tag{11.15}$$

$$f \in C([0,T];V). \tag{11.16}$$

With this notation, a standard calculation based on Green's formula (8.35) leads to the following variational formulation of the frictional problem (11.1)–(11.4).

Problem 11.2. *Find a displacement field $u : [0,T] \to V$ such that*

$$a(u(t), v - u(t)) + \left(\int_0^t A(t-s)\,u(s)\,ds,\ v - u(t)\right)_V + j(v) - j(u(t))$$
$$\ge (f(t), v - u(t))_V \quad \forall\, v \in V,\ t \in [0,T]. \tag{11.17}$$

The difference between the variational model (11.17) and inequality (9.14) deduced in the study of the corresponding elastic problem arises in the fact that (11.17) involves the Volterra operator and the time variable, as well.

In the study of Problem 11.2, we have the following existence and uniqueness result.

Theorem 11.3. *Assume that* (11.5)–(11.8) *hold. Then Problem* 11.2 *has a unique solution $u \in C([0,T];V)$.*

Proof. The bilinear form a is symmetric and, using the condition (11.5), we deduce

$$|a(u,v)| \le \|\mu\|_{L^\infty(\Omega)}\,\|u\|_V\,\|v\|_V \quad \forall\, u,\, v \in V, \tag{11.18}$$

$$a(v,v) \ge \mu^*\,\|v\|_V^2 \quad \forall\, v \in V. \tag{11.19}$$

Also, (11.8) implies that j is a continuous seminorm on V and, therefore, is a convex l.s.c. function. These properties combined with the regularity (11.15) and (11.16) allow us to apply Theorem 6.1 to conclude the proof of Theorem 11.3. □

It follows from Theorem 11.3 that, under the assumptions stated above, the antiplane frictional problem 11.1 has a unique weak solution. We now complete this existence and uniqueness result with the following regularity result.

Theorem 11.4. *Under the conditions stated in Theorem* 11.3, *if there exists* $p \in [1, \infty]$ *such that*

$$f_0 \in W^{1,p}(0,T;L^2(\Omega)), \quad f_2 \in W^{1,p}(0,T;L^2(\Gamma_3)), \tag{11.20}$$

$$\theta \in W^{1,p}(0,T;L^\infty(\Omega)), \tag{11.21}$$

then $u \in W^{1,p}(0,T;V)$.

Proof. We use (11.14) to see that (11.20) implies $f \in W^{1,p}(0,T;V)$. Also, it follows from (11.13) that (11.21) implies $A \in W^{1,p}(0,T;\mathcal{L}(V))$. Theorem 11.4 is now a consequence of Theorem 6.3 in Section 6.1. □

We now consider the static viscoelastic problem with regularized friction. The classical formulation of the problem is as follows.

Problem 11.5. *Find a displacement field* $u_\rho : \Omega \times [0,T] \to \mathbb{R}$ *such that*

$$\operatorname{div}\left(\mu \nabla u_\rho(t) + \int_0^t \theta(t-s)\, \nabla u_\rho(s)\, ds\right) + f_0(t) = 0 \quad \text{in } \Omega, \tag{11.22}$$

$$u_\rho(t) = 0 \quad \text{on } \Gamma_1, \tag{11.23}$$

$$\mu\, \partial_\nu u_\rho(t) + \int_0^t \theta(t-s)\, \partial_\nu u_\rho(s)\, ds = f_2(t) \quad \text{on } \Gamma_2, \tag{11.24}$$

$$\mu\, \partial_\nu u_\rho(t) + \int_0^t \theta(t-s)\, \partial_\nu u_\rho(s)\, ds = -g\, \frac{u_\rho(t)}{\sqrt{u_\rho^2(t) + \rho^2}} \quad \text{on } \Gamma_3, \tag{11.25}$$

for all $t \in [0,T]$.

Note that Problem 11.5 represents a version of the elastic problem 9.4 studied in Section 9.1, stated in the case of viscoelastic materials with long memory.

In the study of Problem 11.5, we assume that (11.5)–(11.8) hold and, moreover, $\rho > 0$. Also, we define the functional $j_\rho : V \to \mathbb{R}$ by

$$j_\rho(v) = \int_{\Gamma_3} g\left(\sqrt{v^2 + \rho^2} - \rho\right) da \quad \forall\, v \in V. \tag{11.26}$$

With this notation, the variational formulation of the antiplane frictional problem 11.5 is the following.

Problem 11.6. *Find a displacement field* $u_\rho : [0,T] \to V$ *such that*

$$a(u_\rho(t), v - u_\rho(t)) + \left(\int_0^t A(t-s)\, u_\rho(s)\, ds,\, v - u_\rho(t)\right)_V + j_\rho(v) - j_\rho(u_\rho(t))$$
$$\geq (f(t), v - u_\rho(t))_V \quad \forall\, v \in V,\ t \in [0,T]. \tag{11.27}$$

Clearly, j_ρ is a convex lower semicontinuous function on V and, moreover, as it was shown in Lemma 9.7, j_ρ and j satisfy condition (3.17). Therefore, similar arguments as those used in the proof of Theorem 11.3, based on the abstract existence and uniqueness result provided by Theorem 6.1 combined with the regularity result in Theorem 6.3 and the abstract convergence result in Theorem 6.10, lead to the following theorem.

Theorem 11.7. *Assume* (11.5)–(11.8). *Then, for every* $\rho > 0$, *Problem* 11.6 *has a unique solution* $u_\rho \in C[0,T];V)$, *and* $u_\rho \in W^{1,p}(0,T;V)$ *if* (11.20) *and* (11.21) *hold. Moreover, the solution converges to the solution* u *of Problem* 10.1, *i.e.,*

$$\|u_\rho - u\|_{C([0,T];V)} \to 0 \quad \text{as } \rho \to 0. \tag{11.28}$$

We conclude by Theorem 11.7 that the antiplane frictional problem (11.22)–(11.25) has a unique weak solution that converges to the weak solution of Problem 11.1 as $\rho \to 0$. In addition to the mathematical interest in this result, it is important from the mechanical point of view since it shows that in the study of static frictional antiplane problems with viscoelastic materials with long memory, the solution of the problem with Tresca's law can be approached as closely as one wishes by the solution of the problem with regularized friction law with a sufficiently small parameter ρ.

11.2 Quasistatic Frictional Problems

In this section, we study the evolutionary versions of the models presented in Section 11.1. The difference arises in the fact that here we replace the static friction laws (11.4) and (11.25) with their evolutionary version (8.62) and (8.72), respectively, and, as a consequence, we need to prescribe an initial condition for the displacement field. We start with the problem involving the Tresca law; its classical formulation is the following.

Problem 11.8. *Find a displacement field* $u : \Omega \times [0,T] \to \mathbb{R}$ *such that*

$$\operatorname{div}\left(\mu\,\nabla u(t) + \int_0^t \theta(t-s)\,\nabla u(s)\,ds\right) + f_0(t) = 0 \quad \text{in } \Omega, \tag{11.29}$$

$$u(t) = 0 \quad \text{on } \Gamma_1, \tag{11.30}$$

$$\mu\,\partial_\nu u(t) + \int_0^t \theta(t-s)\,\partial_\nu u(s)\,ds = f_2(t) \quad \text{on } \Gamma_2, \tag{11.31}$$

$$\left.\begin{aligned} &\left|\mu\partial_\nu u(t) + \int_0^t \theta(t-s)\partial_\nu\, u(s)\,ds\right| \le g, \\ &\mu\,\partial_\nu u(t) + \int_0^t \theta(t-s)\partial_\nu u(s)\,ds \\ &\quad = -g\frac{\dot u(t)}{|\dot u(t)|} \quad \text{if } \dot u(t) \ne 0 \end{aligned}\right\} \quad \text{on } \Gamma_3, \tag{11.32}$$

for all $t \in [0,T]$,

$$u(0) = u_0 \quad \text{in } \Omega. \tag{11.33}$$

In the study of this problem, we assume that the Lamé coefficient μ satisfies (11.5) and the friction bound g satisfies (11.8). We also assume that the relaxation coefficient θ has the regularity

$$\theta \in W^{1,2}(0,T;L^\infty(\Omega)) \tag{11.34}$$

and the body forces and surface tractions densities satisfy

$$f_0 \in W^{1,2}(0,T;L^2(\Omega)), \quad f_2 \in W^{1,2}(0,T;L^2(\Gamma_2)). \tag{11.35}$$

Finally, we assume that the initial data satisfies

$$u_0 \in V, \tag{11.36}$$

$$a(u_0,\, v) + j(v) \geq (f(0),\, v) \quad \forall\, v \in V \tag{11.37}$$

where, here and below, we use the notation (11.9)–(11.12). We note that (11.37) represents a compatibility condition for the initial displacement, already used in Section 9.3 in the study of quasistatic frictional problems with elastic materials.

The variational formulation of Problem 11.8 is the following.

Problem 11.9. *Find a displacement field* $u : [0,T] \to V$ *such that*

$$a(u(t), v - \dot{u}(t)) + \left(\int_0^t A(t-s)\, u(s)\, ds,\, v - \dot{u}(t) \right)_V + j(v) - j(\dot{u}(t))$$

$$\geq (f(t), v - \dot{u}(t))_V \quad \forall\, v \in V,\ \text{a.e. } t \in (0,T), \tag{11.38}$$

$$u(0) = u_0. \tag{11.39}$$

The difference between the variational model (11.38)–(11.39) and the inequality (11.17) deduced in the study of the corresponding static frictional contact problem arises from the fact that in (11.38)–(11.39), we use the test function $v - \dot{u}(t)$, and in (11.17), we use the test function $v - u(t)$, where v is arbitrary in V. As a consequence, in Problem 11.9 we need to prescribe an initial condition, (11.39), and (11.38)–(11.39) represents an evolutionary variational inequality.

In the study of Problem 11.9, we have the following existence and uniqueness result.

Theorem 11.10. *Assume that* (11.5), (11.8), (11.34)–(11.37) *hold. Then there exists a unique solution to Problem* 11.9 *and, moreover, it satisfies* $u \in W^{1,2}(0,T;V)$.

Proof. Using (11.5) and (11.8), it follows that the bilinear form a satisfies assumption (6.2) and the functional j satisfies assumption (6.4). Next, combining (11.13) with (11.34), it follows that $A \in W^{1,2}(0,T;\mathcal{L}(V))$, and combining (11.14) with (11.35), it follows that $f \in W^{1,2}(0,T;V)$; thus, we conclude that the operator A and the function f satisfy conditions (6.14) and (6.15), respectively. Finally, we use (11.36) and (11.37) to see that conditions (6.16) and (6.17) are satisfied, too, the last one with $\delta_0 = 0$. Theorem 11.10 is now a consequence of Theorem 6.4 on page 113. □

We conclude by Theorem 11.10 that the quasistatic frictional problem 11.1 has a unique weak solution.

We consider now the quasistatic viscoelastic problem with regularized friction. The classical formulation of the problem is as follows.

Problem 11.11. *Find a displacement field $u_\rho : \Omega \times [0,T] \to \mathbb{R}$ such that*

$$\operatorname{div}\left(\mu\,\nabla u_\rho(t) + \int_0^t \theta(t-s)\,\nabla u_\rho(s)\,ds\right) + f_0(t) = 0 \quad \text{in } \Omega, \tag{11.40}$$

$$u_\rho(t) = 0 \quad \text{on } \Gamma_1, \tag{11.41}$$

$$\mu\,\partial_\nu u_\rho(t) + \int_0^t \theta(t-s)\,\partial_\nu u_\rho(s)\,ds = f_2(t) \quad \text{on } \Gamma_2, \tag{11.42}$$

$$\mu\,\partial_\nu u_\rho(t) + \int_0^t \theta(t-s)\,\partial_\nu u_\rho(s)\,ds = -g\,\frac{\dot u_\rho(t)}{\sqrt{\dot u_\rho^2(t) + \rho^2}} \quad \text{on } \Gamma_3, \tag{11.43}$$

for all $t \in [0,T]$,

$$u_\rho(0) = u_0 \quad \text{in } \Omega. \tag{11.44}$$

Note that Problem 11.11 represents a version of the elastic problem 9.17 studied in Section 9.1, stated in the case of viscoelastic materials with long memory.

In the study of Problem 11.11, we assume that (11.5), (11.8), (11.34)–(11.37) hold and, moreover, $\rho > 0$. Also, besides the notation already used in Problem 11.9, we use the notation (11.26) for the function $j_\rho : V \to \mathbb{R}$.

The variational formulation of the antiplane frictional Problem 11.11 is the following.

Problem 11.12. *Find a displacement field $u_\rho : [0,T] \to V$ such that*

$$a(u_\rho(t), v - \dot u_\rho(t)) + \left(\int_0^t A(t-s)\,u_\rho(s)\,ds,\, v - \dot u_\rho(t)\right)_V + j_\rho(v) - j_\rho(\dot u_\rho(t))$$
$$\geq (f(t), v - \dot u_\rho(t))_V \quad \forall v \in V, \text{ a.e. } t \in (0,T), \tag{11.45}$$

$$u(0) = u_0. \tag{11.46}$$

Clearly, j_ρ is a convex lower semicontinuous function on V and, moreover, as was shown in the proof of Lemma 9.7, j_ρ and j satisfy condition (3.17). Therefore, similar arguments as those used in the proof of Theorem 11.10, based on the abstract existence and uniqueness result provided by Theorem 6.4 combined with the abstract convergence result in Theorem 6.11, lead to the following theorem.

Theorem 11.13. *Assume* (11.5), (11.8), (11.34)–(11.37). *Then, for every* $\rho > 0$, *Problem* 11.12 *has a unique solution* $u_\rho \in W^{1,2}(0,T;V)$. *Moreover the solution converges to the solution* u *of Problem* 11.8, *i.e.,*

$$\|u_\rho - u\|_{C([0,T];V)} \to 0 \quad \text{as } \rho \to 0. \tag{11.47}$$

We conclude by Theorem 11.13 that the antiplane frictional problem (11.40)–(11.44) has a unique weak solution that converges to the weak solution of Problem 11.8 as $\rho \to 0$. In addition to the mathematical interest in this result, it is important from the mechanical point of view since it shows that, in the study of quasistatic antiplane contact problems for viscoelastic materials with long memory, the solution of the problem with Tresca's law can be approached as closely as one wishes by the solution of the contact problem with regularized friction law with a sufficiently small parameter ρ.

11.3 Approach to Elasticity

In this section, we investigate the behavior of the weak solutions of the antiplane frictional problems for viscoelastic materials with long memory as the relaxation coefficient converges to zero. We consider both the static and the quasistatic cases. Our results below are valid both for the Tresca friction law and its regularization; however, to avoid repetition, we restrict ourselves to the case of the Tresca friction law.

We start with the static case and, in order to outline the dependence of the solution on the given relaxation coefficient θ, we rewrite Problem 11.2 as follows.

Problem 11.14. *Find a displacement field* $u_\theta : [0,T] \to V$ *such that*

$$a(u_\theta(t), v - u_\theta(t)) + \left(\int_0^t A_\theta(t-s)\, u_\theta(s)\, ds,\ v - u_\theta(t) \right)_V + j(v) - j(u_\theta(t))$$
$$\geq (f(t), v - u_\theta(t))_V \quad \forall\, v \in V,\ t \in [0,T]. \tag{11.48}$$

Here the operator A_θ is given by (11.11), i.e.,

$$(A_\theta(t)\, u,\, v)_V = \int_\Omega \theta(t)\, \nabla u \cdot \nabla v\, dx \quad \forall\, u, v \in V,\ t \in [0,T]. \tag{11.49}$$

We also consider the problem obtained for $\theta = 0$, that is:

Problem 11.15. *Find a displacement field $u : [0,T] \to V$ such that*

$$a(u(t), v - u(t)) + j(v) - j(u(t)) \geq (f(t), v - u(t))_V \quad \forall v \in V,\ t \in [0,T]. \tag{11.50}$$

It is easy to see that Problem 11.15 represents a version of the elastic problem 9.2, obtained in the case when the body forces and tractions depend on time, and is formulated as a time-dependent variational inequality.

We assume in what follows that (11.5)–(11.8) hold. It follows from Theorem 11.3 that Problem 11.14 has a unique solution $u_\theta \in C([0,T];V)$, and it follows from Theorem 3.12 that Problem 11.15 has a unique solution $u \in C([0,T];V)$. Consider now the assumption

$$\|\theta\|_{C([0,T];L^\infty(\Omega))} \to 0. \tag{11.51}$$

We have the following convergence result.

Theorem 11.16. *Assume that* (11.5)–(11.8) *and* (11.51) *hold. Then the solution u_θ of Problem* 11.14 *converges to the solution u of Problem* 11.15, *i.e.,*

$$\|u_\theta - u\|_{C([0,T];V)} \to 0. \tag{11.52}$$

Proof. It follows from (11.49) that

$$\|A_\theta(t)\|_{\mathcal{L}(V)} \leq \|\theta(t)\|_{L^\infty(\Omega)} \quad \forall t \in [0,T]$$

and, therefore,

$$\|A_\theta\|_{C([0,T];\mathcal{L}(V))} \leq \|\theta\|_{C([0,T];L^\infty(\Omega))}. \tag{11.53}$$

Assumption (11.51) combined with inequality (11.53) yields

$$\|A_\theta\|_{C([0,T];\mathcal{L}(V))} \to 0. \tag{11.54}$$

Note that the solution u of Problem 11.15 satisfies a variational inequality of the form (6.1) with $A = 0_{\mathcal{L}(V)}$ and, therefore, (11.54) implies (6.33). The convergence result (11.52) is now a consequence of Theorem 6.8. □

We turn now to the quasistatic case. Again, in order to outline the dependence of various functions with respect to θ, we reformulate the Problem 11.9 as follows.

Problem 11.17. *Find a displacement field $u_\theta : [0,T] \to V$ such that*

$$a(u_\theta(t), v - \dot{u}_\theta(t)) + \Big(\int_0^t A_\theta(t-s)\, u_\theta(s)\, ds,\, v - \dot{u}_\theta(t) \Big)_V + j(v) - j(\dot{u}_\theta(t))$$
$$\geq (f(t), v - \dot{u}_\theta(t))_V \quad \forall\, v \in V, \text{ a.e. } t \in (0,T), \tag{11.55}$$
$$u_\theta(0) = u_0. \tag{11.56}$$

We also consider the problem obtained for $\theta = 0$, that is:

Problem 11.18. *Find a displacement field $u : [0,T] \to V$ such that*

$$a(u(t), v - \dot{u}(t)) + j(v) - j(\dot{u}(t))$$
$$\geq (f(t), v - \dot{u}(t))_V \quad \forall\, v \in V, \text{ a.e. } t \in (0,T), \tag{11.57}$$
$$u(0) = u_0. \tag{11.58}$$

Clearly, Problem 11.18 is the elastic problem 9.16 studied in Section 9.3.

We assume in what follows that (11.5), (11.8), and (11.34)–(11.37) hold. It follows from Theorem 11.10 that Problem 11.17 has a unique solution $u_\theta \in W^{1,2}(0,T;V)$, and it follows from Theorem 9.19(2) that Problem 11.18 has a unique solution $u \in W^{1,2}(0,T;V)$.

Consider now the assumption

$$\|\theta\|_{W^{1,2}(0,T;L^\infty(\Omega))} \to 0. \tag{11.59}$$

We have the following convergence result.

Theorem 11.19. *Assume that* (11.5), (11.8), (11.34)–(11.37), *and* (11.59) *hold. Then the solution u_θ of Problem* 11.17 *converges to the solution u of Problem* 11.18, *i.e.,*

$$\|u_\theta - u\|_{C([0,T];V)} \to 0. \tag{11.60}$$

Proof. It follows from (11.49) that

$$\|A_\theta(t)\|_{\mathcal{L}(V)} \leq \|\theta(t)\|_{L^\infty(\Omega)} \quad \forall t \in [0,T]$$

and

$$\|A_\theta(t_1) - A_\theta(t_2)\|_{\mathcal{L}(V)} \leq \|\theta(t_1) - \theta(t_2)\|_{L^\infty(\Omega)} \quad \forall t_1,\, t_2 \in [0,T]$$

which imply that

$$\|A_\theta\|_{W^{1,2}(0,T;\mathcal{L}(V))} \leq \|\theta\|_{W^{1,2}(0,T;V)}. \tag{11.61}$$

Assumption (11.59) combined with inequality (11.61) yields

$$\|A_\theta\|_{W^{1,2}(0,T;\mathcal{L}(V))} \to 0. \tag{11.62}$$

Note that the solution u of Problem 11.18 satisfies a variational inequality of the form (6.12) with $A = 0_{\mathcal{L}(V)}$ and, therefore, (11.62) implies (6.44). The convergence result (11.60) is now a consequence of Theorem 6.9. □

From the convergence results in Theorems 11.16 and 11.19, we conclude that, in the study of antiplane contact problems with Tresca's friction law, the solution of the viscoelastic problem with long memory may be approached by the solution of the corresponding elastic problem as the relaxation function is small enough. This approach holds both in the static and quasistatic cases.

Bibliographical Notes

In writing Section 9.1, we followed some ideas of [53], where antiplane static frictional contact problems for elastic materials were considered, with the emphasis on their numerical analysis; there, the contact was modelized with Tresca's friction law (9.4) and with its regularizations (9.19) and (9.36). Various static frictional contact problems for nonlinear elastic materials were studied in [5], in the antiplane context; there, the concept of entropy solutions for such kind of problems was introduced and existence results were proved. A frictional antiplane contact problem for elastic materials with Coulomb's law was considered in [50] and a static slip-dependent frictional problem for nonlinear elastic materials was studied in [27], in the three-dimensional case; in this last reference the existence of a unique solution of the problem was obtained by using an abstract result for elliptic quasivariational inequalities, proved in [107].

The first mathematical results in the study of Problem 9.11 were obtained in [79]; there, the authors prove the existence of a solution to the problem by using a Weierstrass type minimization theorem, then give sufficient conditions for uniqueness and stability; moreover, they analyze the bifurcation points between different branches of solutions. In writing Section 9.2 we used [27], since the problem studied in this section represents the antiplane version of the three-dimensional problem studied in this last reference.

Section 9.3 follows some ideas presented in [37]. There, the solvability of the quasistatic frictional contact problems for elastic-viscoplastic materials was obtained via a regularization method; two regularizations of the Tresca friction law were considered, (7.50) and (7.51), together with their corresponding discrete scheme; convergence results were proved that show that considering the regularized friction law (7.50) provides a better approximation of the solution than considering the regularized friction law (7.51); numerical simulations illustrating this conclusion were also presented. The numerical analysis of a quasistatic frictional contact problem of the form (9.65)–(9.66), including error estimates for the discrete solution and numerical simulations, was studied in [10].

Section 9.4 follows, with some modifications, the results obtained in [100, 103]. Note also that a three-dimensional slip-dependent quasistatic frictional problem that leads to an evolutionary quasivariational inequality of the form (9.85)–(9.86) was studied in [33]. There, besides the existence and uniqueness of the solution in displacements, a dual formulation in stress was considered, and the relationship between the primal and the dual problem was investigated.

The convergence result in Theorem 10.10 was obtained in [68] in the homogeneous case $\theta(\boldsymbol{x}) = \theta$. A version of this theorem was obtained in [132] in the case of antiplane problems with power-law friction (8.73), (8.74). A similar convergence result was obtained in [25] in the study of three-dimensional frictional contact problems. The results presented in Section 10.3 represent the antiplane version of the results obtained in [3] in the study of three-dimensional slip- and slip rate–dependent frictional problems for viscoelastic materials. The study of the total slip rate–dependent frictional problems presented in Section 10.4 was performed in [68, 100, 101]. There, the existence and uniqueness result in Theorem 10.20 was proved by using the Banach fixed point theorem, twice.

Quasistatic antiplane problems for viscoelastic materials with long memory were considered in [102, 135], where Theorems 11.10, 11.13, and 11.19 were proved. The rest of the results in Chapter 11 represents antiplane versions of some results obtained in [123].

We now provide additional comments and references on antiplane shear deformations, including antiplane frictional contact problems.

First, we note that antiplane frictional contact problems were used in geophysics in order to describe pre-earthquake evolution of the regions of hight tectonic activity. Indeed, understanding the frictional behavior of a fault is a key issue of earthquake physics. References on this topic include [19, 20, 83, 115], where the physical models are described and numerical results are presented, together with various comments and conclusions. The mathematical analysis of the problem of antiplane shearing on a periodic system of faults under a slip-dependent friction law in linear elastodynamics was performed in [74, 81]; there, a spectral analysis was performed in order to characterize the existence of unstable solutions. In all these papers, the models presented were formulated in the context of pre-stressed reference configurations described in Section 8.5.

A model for dynamic contact processes involving elastic materials in the context of antiplane deformations was derived in [117]; there, the coefficient of friction was assumed to depend on the tangential velocity, and this dependence was explained by the form of the stress in the asperities as a function of time. The mathematical analysis of dynamic antiplane contact problems with slip-dependent friction was provided in [77, 78]; there, it was proved that the solution is not uniquely determined and presents shocks.

A dynamic contact problem with slip-dependent coefficient of friction for linearly elastic materials was considered in [76], under the assumption that the normal stress is prescribed on the contact surface, i.e., it satisfies (7.30). In a weak formulation, this problem leads to a second-order hyperbolic variational inequality. The existence of the solution was obtained in the two-dimensional case and in the antiplane context, as well. The proof, based on the Galerkin method, is classical and constructive; it uses regularization techniques similar to those presented in this book and compactness arguments. The corresponding two-dimensional dynamic viscoelastic frictional contact problem was also considered, and it was proved that its solution converges to the solution of the two-dimensional dynamic elastic problem, as the viscosity coefficient converges to zero.

An efficient nonconforming discretization method for the frictional contact between two elastic bodies was developed in [73] and tested in the context of antiplane shear deformations. There, the method was based on a mixed variational formulation in which dual basis functions were used to discretize the Lagrange multiplier and an optimal a priori error estimate was obtained. The performance of the algorithm was illustrated by numerical results.

A mathematical problem that describes the deformation of a linearly elastic body in adhesive contact was considered in [134], in the context of antiplane shear deformations. There, the unique solvability of the problem was obtained under a suitable smallness assumption on the problem data. The study of the adhesive problem in this simple setting allowed the study of the adhesive contact in the general three-dimensional case to proceed as shown in the recent monograph [133] and the references therein.

The effect of mechanical/electric coupling on the decay of Saint-Venant end effects in linear piezoelectricity was investigated in [14] in the context of antiplane shear deformations for linear piezoelectric solids. The current rapidly developing smart structures technology provides motivation for the investigation of such problems, see for instance [157] where the behavior of two collinear symmetric cracks subjected to the antiplane shear loading in piezoelectric materials was investigated.

References

1. R.A. Adams, *Sobolev Spaces*, Academic Press, New York, 1975.
2. A. Amassad, C. Fabre and M. Sofonea, A Quasistatic Viscoplastic Contact Problem with Normal Compliance and Friction, *IMA Journal of Applied Mathematics* **69** (2004), 463–482.
3. A. Amassad, M. Shillor and M. Sofonea, A quasistatic contact problem with slip dependent coefficient of friction, *Math. Meth. Appl. Sci.* **22** (1999), 267–284.
4. A. Amassad and M. Sofonea, Analysis of a quasistatic viscoplastic problem involving Tresca friction law, *Discrete and Continuous Dynamical Systems* **4** (1998), 55–72.
5. F. Andreu, J. M. Mazón and M. Sofonea, Entropy solutions in the study of antiplane shear deformations for elastic solids, *Mathematical Models and Methods in Applied Sciences* (M^3AS) **10** (2000), 96–126.
6. S.S. Antman, *Nonlinear Problems of Elasticity*, Springer-Verlag, New York, 1995.
7. K. Atkinson and W. Han, *Theoretical Numerical Analysis: A Functional Analysis Framework*, Texts in Applied Mathematics **39**, Springer, New York, 2001.
8. B. Awbi, M. Rochdi and M. Sofonea, Abstract evolution equations for viscoelastic frictional contact problems, *Journal of Applied Mathematics and Physics* (*ZAMP*) **50** (1999), 1–18.
9. C. Baiocchi and A. Capelo, *Variational and Quasivariational Inequalities: Applications to Free-Boundary Problems*, John Wiley, Chichester, 1984.
10. M. Barboteu, W. Han and M. Sofonea, Numerical analysis of a bilateral frictional contact problem for linearly elastic materials, *IMA Journal of Numerical Analysis* **22** (2002), 407–436.
11. V. Barbu, *Nonlinear Semigroups and Differential Equations in Banach Spaces*, Editura Academiei, Bucharest-Noordhoff, Leyden, 1976.
12. V. Barbu, *Optimal Control of Variational Inequalities*, Pitman, Boston, 1984.
13. V. Barbu, T. Precupanu, *Convexity and Optimization in Banach Spaces*, D. Reidel Publishing Company, Dordrecht, 1986.
14. A. Borrelli, C. O. Horgan and M. C. Patria, Saint-Venant's principle for antiplane shear deformations of linear piezoelectric materials, *SIAM J. Appl. Math.* **62** (2002), 2027–2044.
15. H. Brézis, Equations et inéquations non linéaires dans les espaces vectoriels en dualité, *Ann. Inst. Fourier* **18** (1968), 115–175.
16. H. Brézis, Problèmes unilatéraux, *J. Math. Pures et Appl.* **51** (1972), 1–168.
17. H. Brézis, *Opérateurs maximaux monotones et semi-groupes de contractions dans les espaces de Hilbert*, Mathematics Studies, North Holland, Amsterdam, 1973.

18. H. Brézis, *Analyse fonctionnelle—Théorie et applications*, Masson, Paris, 1987.
19. M. Campillo, C. Dascalu and I.R. Ionescu, Instability of a periodic system of Faults, *Geophysical International Journal* **159** (2004), 212–222.
20. M. Campillo and I.R. Ionescu, Initiation of antiplane shear instability under slip dependent friction, *Journal of Geophysical Research* **102 B9** (1997) 363–371.
21. M. M. Caroll, Some results on finite amplitude elastic waves, *Acta Mechanica*, **3** (1967) 167–181.
22. T. Cazenave, A. Haraux, *Introduction aux problèmes d'évolution semi-linéaires*, Ellipses, Paris, 1990.
23. Y.-S. Chan, G.H. Paulino and A.C. Fannjiang, The crack problem for nonhomogeneous materials under antiplane shear loading—a displacement based formulation, *International Journal of Solids and Structures* **38** (2001), 2989–3005.
24. O. Chau, W. Han and M. Sofonea, Analysis and approximation of a viscoelastic contact problem with slip dependent friction, *Dynamics of Continuous, Discrete and Impulsive Systems* **8** (2001), 153–174.
25. O. Chau, D. Motreanu and M. Sofonea, Quasistatic frictional problems for elastic and viscoelastic materials, *Applications of Mathematics* **47** (2002), 341–360.
26. P.G. Ciarlet, *Mathematical Elasticity, Volume I: Three Dimensional Elasticity*, Studies in Mathematics and its Applications, Vol. 20, North-Holland, Amsterdam, 1988.
27. C. Ciulcu, D. Motreanu and M. Sofonea, Analysis of an elastic contact problem with slip dependent coefficient of friction, *Mathematical Inequalities & Applications* **4** (2001), 465–479.
28. M. Cocu, Existence of solutions of Signorini problems with friction, *Int. J. Eng. Sci.* **22** (1984), 567–581.
29. M. Cocu, E. Pratt and M. Raous, Existence d'une solution du problème quasistatique de contact unilatéral avec frottement non local, *C. R. Acad. Sci. Paris,* **320**, Série I (1995), 1413–1417.
30. M. Cocu, E. Pratt and M. Raous, Formulation and approximation of quasistatic frictional contact, *Int. J. Eng. Sci.* **34** (1996), 783–798.
31. M. Cocu and J.M. Ricaud, Existence results for a class of implicit evolution inequalities and application to dynamic unilateral contact problems with friction, *C. R. Acad. Sci. Paris* **329**, Série I (1999), 839–844.
32. M. Cocu and J.M. Ricaud, Analysis of a class of implicit evolution inequalities associated to dynamic contact problems with friction, *Int. J. Eng. Sci.* **328** (2000), 1534–1549.
33. C. Corneschi, T.-V. Hoarau-Mantel and M. Sofonea, A quasistatic contact problem with slip dependent coefficient of friction for elastic materials, *Journal of Applied Analysis* **8** (2002), 59–80.
34. N. Cristescu, *Rocky Rheology*, Kluwer Academic Publishers, Dordrecht, 1989.
35. N. Cristescu, I.R. Ionescu and I. Rosca, A numerical analysis of the foot-floor interaction in long wall workings, *International Journal for Numerical and Analytical Methods in Geomechanics* **18** (1994), 641–652.
36. N. Cristescu and I. Suliciu, *Viscoplasticity*, Martinus Nijhoff Publishers, Editura Tehnică, Bucharest, 1982.
37. M. Delost, *Analyse théorique et numérique pour des problèmes quasistatiques régularisés de contact avec frottement*, Ph. D. Thesis, Université de Nice-Sophia Antipolis, Nice, 2004.
38. Z. Denkowski, S. Migórski and N. S. Papageorgiu, *An Introduction to Nonlinear Analysis: Theory*, Kluwer Academic/Plenum Publishers, Boston, 2003.
39. Z. Denkowski, S. Migórski and N. S. Papageorgiu, *An Introduction to Nonlinear Analysis: Applications*, Kluwer Academic/Plenum Publishers, Boston, 2003.
40. I. Doghri, *Mechanics of Deformable Solids*, Springer, Berlin, 2000.

41. A.D. Drozdov, *Finite Elasticity and Viscoelasticity—A Course in the Nonlinear Mechanics of Solids*, World Scientific, Singapore, 1996.
42. G. Duvaut and J.-L. Lions, *Inequalities in Mechanics and Physics*, Springer-Verlag, Berlin, 1976.
43. C. Eck, J. Jarušek and M. Krbeč, *Unilateral Contact Problems: Variational Methods and Existence Theorems*, Pure and Applied Mathematics **270**, Chapman/CRC Press, New York, 2005.
44. I. Ekeland and R. Temam, *Convex Analysis and Variational Problems*, North-Holland, Amsterdam, 1976.
45. H. Engler, Global regular solutions of the dynamic antiplane shear problem in nonlinear viscoelasticity, *Math. Zeit.*, A **202** (1989) 251–259.
46. L.C. Evans, *Partial Differential Equations*, AMS Press, Providence, 1999.
47. G. Fichera, Problemi elastostatici con vincoli unilaterali. II. Problema di Signorini con ambique condizioni al contorno, *Mem. Accad. Naz. Lincei, S. VIII, Vol. VII, Sez. I*, **5** (1964) 91–140.
48. R.L. Fosdick and B. Kao, Transverse deformations associated with rectilinear shear in elastic solids, *Journal of Elasticity* **8** (1978) 117-142.
49. A. Friedman, *Variational Principles and Free-boundary Problems*, John Wiley, New York, 1982.
50. B.Z. Gai, Frictional slip between a gradient non-homogenuous layer and a half-space in antiplane elastic wave field, *Extended Summaries of the 21 st International Congress of Theoretical and Applied Mechanics (ICTAM 04)*, Warsaw, 2004, CD-ROM.
51. P. Germain and P. Muller, *Introduction à la mécanique des milieux continus*, Masson, Paris, 1980.
52. R. Glowinski, *Numerical Methods for Nonlinear Variational Problems*, Springer-Verlag, New York, 1984.
53. R. Glowinski, J.-L. Lions and R. Trémolières, *Numerical Analysis of Variational Inequalities*, North-Holland, Amsterdam, 1981.
54. J. M. Greenberg and A. Nouri, Antiplane shearing motions of a visco-plastic solid, *SIAM Journal of Mathematical Analysis* **24** (1993), 943–967.
55. A. Guran, F. Pfeiffer and K. Popp, eds., *Dynamics with Friction: Modeling, Analysis and Experiment, Part I*, World Scientific, Singapore, 1996.
56. M.E. Gurtin, *An Introduction to Continuum Mechanics*, Academic Press, New York, 1981.
57. W. Han and B.D. Reddy, *Plasticity: Mathematical Theory and Numerical Analysis*, Springer-Verlag, New York, 1999.
58. W. Han and M. Sofonea, Evolutionary variational inequalities arising in viscoelastic contact problems, *SIAM Journal of Numerical Analysis* **38** (2000), 556–579.
59. W. Han and M. Sofonea, Time-dependent variational inequalities for viscoelastic contact problems, *Journal of Computational and Applied Mathematics* **136** (2001), 369–387.
60. W. Han and M. Sofonea, *Quasistatic Contact Problems in Viscoelasticity and Viscoplasticity*, Studies in Advanced Mathematics **30**, American Mathematical Society, Providence, RI—International Press, Somerville, MA, 2002.
61. J. Haslinger, I. Hlaváček and J. Nečas, Numerical methods for unilateral problems in solid mechanics, *Handbook of Numerical Analysis, Vol IV*, P.G. Ciarlet and J.-L. Lions, eds., North-Holland, Amsterdam, 1996, 313–485.
62. M.A. Hayes and K.R. Rajagopal, Inhomogeneous finite amplitude motions in a neo-Hookean solid, *Proc. R. Irish Acad.*, A **92** (1992) 137–147.
63. M.A. Hayes and G. Saccomandi, Anti-plane shear motions for viscoelastic Mooney-Rivlin materials, *Q. Jl. Mech. Appl. Math.* **57** (2004), 379–392.
64. R. Hertzberg, *Deformation and Fracture Mechanics of Engineering Materials*, John Wiley and Sons, New York, 1996.

65. P. Hild, On finite element uniqueness studies for Coulomb's frictional contact model, *Int. J. Appl. Math. Comput. Sci.* **12** (2002), 41–50.
66. J.-B. Hiriart-Urruty and C. Lemaréchal, *Convex Analysis and Minimization Algorithms, I, II*, Springer-Verlag, Berlin, 1993.
67. I. Hlaváček, J. Haslinger, J. Nečas and J. Lovíšek, *Solution of Variational Inequalities in Mechanics*, Springer-Verlag, New York, 1988.
68. T.-V. Hoarau-Mantel and A. Matei, Analysis of a viscoelastic antiplane contact problem with slip-dependent friction, *International Journal of Applied Mathematics and Computer Science* **12** (2002), 51–58.
69. C.O. Horgan, Anti-plane shear deformation in linear and nonlinear solid mechanics, *SIAM Rev.* **37** (1995), 53–81.
70. C.O. Horgan, Recent developments concerning Saint-Venant's principle: a second update, *Appl. Mech. Rev.* **49** (1996), S101–S111.
71. C.O. Horgan and K. L. Miller, Anti-plane shear deformation for homogeneous and inhomogeneous anisotropic linearly elastic solids, *J. Appl. Mech.* **61** (1994), 23-29.
72. C.O. Horgan and G. Saccomandi, Superposition of generalized plane strain on anti-plane shear deformation in isotropic incompressible hyperelastic materials, *Journal of Elasticity* **73** (2003), 221–235.
73. S. Hüeber, A. Matei and B.I. Wohlmuth, Efficient algorithms for problems with friction, *SIAM Journal on Scientific Computing* **29** (2007), 70–92.
74. I.R. Ionescu, C. Dascalu and M. Campillo, Slip-weakening friction on a periodic System of faults: spectral analysis, *Z. Angew. Math. Phys. (ZAMP)* **53** (2002), 980–995.
75. I.R. Ionescu and Q.-L. Nguyen, Dynamic contact problems with slip dependent friction in viscoelasticity, *Int. J. Appl. Math. Comput. Sci.* **12** (2002), 71–80.
76. I.R. Ionescu, Q.-L. Nguyen and S. Wolf, Slip displacement dependent friction in dynamic elasticity, *Nonlinear Analysis* **53** (2003), 375–390.
77. I.R. Ionescu and J.-C. Paumier, Friction dynamique avec coefficient dépendant de la vitesse de glissement, *C. R. Acad. Sci. Paris* **316**, Série I (1993), 121–125.
78. I.R. Ionescu and J.-C. Paumier, On the contact problem with slip rate dependent friction in elastodynamics, *European J. Mech., A—Solids* **13** (1994), 556–568.
79. I.R. Ionescu and J.-C. Paumier, On the contact problem with slip displacement dependent friction in elastostatics, *Int. J. Eng. Sci.* **34** (1996), 471–491.
80. I.R. Ionescu and M. Sofonea, *Functional and Numerical Methods in Viscoplasticity*, Oxford University Press, Oxford, 1993.
81. I.R. Ionescu and S. Wolf, Interaction of faults under slip dependent friction. Nonlinear eingenvalue analysis, *Mathematical Methods in Applied Sciences* (M^2AS) **28** (2005), 77–100.
82. K.L. Johnson, *Contact Mechanics*, Cambridge University Press, Cambridge, 1987.
83. D. Kiyashchenko and V. Troyan, The pecularities of shear crack pre-rupture evolution and distribution of seismicity before strong earthquakes, *Natural Hazards and Earth System Sciences* **1** (2001), 145–158.
84. A.M. Khludnev and J. Sokolowski, *Modelling and Control in Solid Mechanics*, Birkhäuser-Verlag, Basel, 1997.
85. N. Kikuchi and J.T. Oden, Theory of variational inequalities with applications to problems of flow through porous media, *Int. J. Eng. Sci.* **18** (1980), 1173–1284.
86. N. Kikuchi and J.T. Oden, *Contact Problems in Elasticity: A Study of Variational Inequalities and Finite Element Methods*, SIAM, Philadelphia, 1988.

87. D. Kinderlehrer and G. Stampacchia, *An Introduction to Variational Inequalities and their Applications*, Classics in Applied Mathematics **31**, SIAM, Philadelphia, 2000.
88. A. Klarbring, A. Mikelič and M. Shillor, Frictional contact problems with normal compliance, *Int. J. Eng. Sci.* **26** (1988), 811–832.
89. A. Klarbring, A. Mikelič and M. Shillor, On friction problems with normal compliance, *Nonlinear Analysis* **13** (1989), 935–955.
90. J.K. Knowles, On finite anti-plane shear for incompressible elastic materials, *Journal of Australian Math. Soc. B* **19** (1976), 400–415.
91. J.K. Knowles, The finite anti-plane shear field near the tip of a crack for a class of incompressible elastic solids, *Internat. J. Fracture* **13** (1977), 611–639.
92. Y. Komura, Nonlinear semigroups in Hilbert Spaces, *Journal of the Mathematical Society of Japan* **19** (1967), 493–507.
93. T.A. Laursen, *Computational Contact and Impact Mechanics*, Springer, Berlin, 2002.
94. J. Lemaître and J.-L. Chaboche, *Mechanics of Solids Materials*, Cambridge University Press, Cambridge, 1990.
95. J.-L. Lions and E. Magenes, *Problèmes aux limites non-homogènes I*, Dunod, Paris, 1968.
96. L.E. Malvern, *Introduction to the Mechanics of a Continuum Medium*, Princeton-Hall, Inc, New Jersey, 1969.
97. F. Martinez, M. Sofonea and J.M. Ségura, Analysis and Numerical Computation in the Study of Pre-stressed Composite Assemblies, *Rev. Roum. Sci. Tech.-Méc. Appl.* **46** (2001), 53–74.
98. J.A.C. Martins and M.D.P. Monteiro Marques, eds., *Contact Mechanics*, Kluwer, Dordrecht, 2002.
99. J.A.C. Martins and J.T. Oden, Existence and uniqueness results for dynamic contact problems with nonlinear normal and friction interface laws, *Nonlinear Analysis TMA* **11** (1987), 407–428.
100. A. Matei, *Modélisation Mathématique en Mécanique du Contact*, Ph. D. Thesis, Université de Perpignan, Perpignan, 2002.
101. A. Matei, Antiplane contact problems for viscoelastic materials, *Annals of University of Craiova, Math. Comp. Sci. Ser.* **30** (2003), 169–176.
102. A. Matei and T.-V. Hoarau-Mantel, Problèmes antiplans de contact avec frottement pour les matériaux viscoélastiques à mémoire longue, *Annals of University of Craiova, Math. Comp. Sci. Ser.* **32** (2005), 200–206.
103. A. Matei, V.V. Motreanu and M. Sofonea, A quasistatic antiplane contact problem with slip dependent friction, *Advances in Nonlinear Variational Inequalities* **4** (2001), 1–21.
104. D. Maugis, *Contact, Adhesion and Rupture of Elastic Solids*, Springer-Verlag, Berlin, Heidelberg, 2000.
105. J.J. Moreau, Proximité et dualité dans un espace hilbertien, *Bulletin de la Société Mathématique de France* **93** (1965), 273–283.
106. D. Motreanu and M. Sofonea, Evolutionary variational inequalities arising in quasistatic frictional contact problems for elastic materials, *Abstract and Applied Analysis* **4** (1999), 255–279.
107. D. Motreanu and M. Sofonea, Quasivariational inequalities and applications in frictional contact problems with normal compliance, *Adv. Math. Sci. Appl.* **10** (2000), 103–118.
108. J. Nečas, *Les méthodes directes en théorie des équations elliptiques*, Academia, Praha, 1967.
109. J. Nečas and I. Hlaváček, *Mathematical Theory of Elastic and Elastico-Plastic Bodies: An Introduction*, Elsevier Scientific Publishing Company, Amsterdam, Oxford, New York, 1981.

110. Q.S. Nguyen, *Stability and Nonlinear Solid Mechanics*, John Wiley & Sons, LTD, Chichester, 2000.
111. J.T. Oden and J.A.C. Martins, Models and computational methods for dynamic friction phenomena, *Computer Methods in Applied Mechanics and Engineering* **52** (1985), 527–634.
112. J.T. Oden and E.B. Pires, Nonlocal and nonlinear friction laws and variational principles for contact problems in elasticity, *Journal of Applied Mechanics* **50** (1983), 67–76.
113. P.D. Panagiotopoulos, *Inequality Problems in Mechanics and Applications*, Birkhäuser, Boston, 1985.
114. G.H. Paulino and Z.-H. Jin, Viscoelastic functionally graded materials subjected to antiplane shear fracture, *Journal of Applied Mechanics* **68** (2001), 284–293.
115. H. Perfettini, M. Campillo and I. R. Ionescu, Rescaling of the weakening rate, *Journal of Geophysical Research* **108** B9 (2003), 2410–2414.
116. A.C. Pipkin, *Lectures in Viscoelasticity Theory*, Applied Mathematical Sciences **7**, George Allen & Unwin Ltd., London, Springer-Verlag, New York, 1972.
117. D. Pisarenko, Elastodynamical mechanism of rate-dependent friction, *Geophysical Journal International* **148** (2002), 499–505.
118. P. Quignon, Contraintes d'interface d'un bi-couche précontraint, *Revue des Composites et des Matériaux Avancés* **4** (1994), 213–240.
119. E. Rabinowicz, *Friction and Wear of Materials*, 2nd edition, Wiley, New York, 1995.
120. M. Raous, M. Jean and J.J. Moreau, eds., *Contact Mechanics*, Plenum Press, New York, 1995.
121. B.D. Reddy, *Introductory Functional Analysis with Applications to Boundary Value Problems and Finite Elements*, Springer, New York, 1998.
122. M. Rochdi, M. Shillor and M. Sofonea, Quasistatic viscoelastic contact with normal compliance and friction, *Journal of Elasticity* **51** (1998), 105–126.
123. A. D. Rodríguez–Aros, M. Sofonea and J. M. Viaño, A Class of Evolutionary Variational Inequalities with Volterra-type Integral Term, *Mathematical Models and Methods in Applied Sciences* (M^3AS) **14** (2004), 555–577.
124. A. D. Rodríguez–Aros, M. Sofonea and J. M. Viaño, Numerical Approximation of a Viscoelastic Frictional Contact Problem, *C. R. Acad. Sci. Paris, Série II Méc.* **334** (2006), 279–284.
125. A. D. Rodríguez–Aros, M. Sofonea and J. M. Viaño, Numerical Analysis of a Frictional Contact Problem for Viscoelastic Materials with Long-term Memory, *Numerische Mathematik* **198** (2007), 327–358.
126. I. Rosca, Functional framework for linear variational equations in *Curent Topics in Continuum Mechanics*, L. Dragos Ed., Editura Academiei Române, Bucharest 2002, 177–258.
127. J.M. Segura and G. Armengaud, Etude des contraintes d'interface dans un assemblage composite constitué de deux plaques précontraintes ou de deux plaques chauffées, *Mechanics Research Communications* **24** (1997), 489–501.
128. M. Shillor, ed., Recent advances in contact mechanics, Special issue of *Math. Comput. Modelling* **28** (4–8) (1998).
129. M. Shillor and M. Sofonea, A quasistatic viscoelastic contact problem with friction, *Int. J. Eng. Sci.* **38** (2000), 1517–1533.
130. M. Shillor, M. Sofonea and J.J. Telega, *Models and Analysis of Quasistatic Contact*, Lecture Notes in Physics **655**, Springer, Berlin, 2004.
131. A. Signorini, Sopra alcune questioni di elastostatica, *Atti della Società Italiana per il Progresso delle Scienze*, 1933.

132. M. Sofonea and M. Ait Mansour, A convergence result for evolutionary variational inequalities and applications to antiplane frictional contact problems, *Applicaciones Mathematicae* **31** (2004), 55–67.
133. M. Sofonea, W. Han and M. Shillor, *Analysis and Approximation of Contact Problems with Adhesion or Damage*, Pure and Applied Mathematics **276**, Chapman-Hall/CRC Press, New York, 2006.
134. M. Sofonea and A. Matei, Elastic antiplane contact problem with Adhesion, *Journal of Applied Mathematics and Physics* (*ZAMP*) **53** (2002), 962–972.
135. M. Sofonea, C. Niculescu and A. Matei, An antiplane contact problem for viscoelastic materials with long-term memory, *Mathematical Modelling and Analysis* **11** (2006), 212–228.
136. M. Sofonea, A. D. Rodríguez–Aros and J. M. Viaño, A class of integro-differential variational inequalities with applications to viscoelastic contact, *Mathematical and Computer Modelling* **41** (2005), 1355–1369.
137. C.H. Scholz, *The Mechanics of Earthquakes and Faulting*, Cambridge University Press, Cambridge, 1990.
138. I. Stakgold, *Green's Functions and Boundary Value Problems*, John Wiley & Sons, Inc., 1979.
139. N. Strömberg, *Thermomechanical Modelling of Tribological Systems*, Ph.D. Thesis, no. 497, Linköping University, Sweden, 1997.
140. N. Strömberg, L. Johansson and A. Klarbring, Generalized standard model for contact friction and wear, in *Contact Mechanics*, M. Raous, M. Jean and J.J. Moreau, eds., Plenum Press, New York, 1995.
141. N. Strömberg, L. Johansson and A. Klarbring, Derivation and analysis of a generalized standard model for contact friction and wear, *Int. J. Solids Structures* **33** (1996), 1817–1836.
142. J.J. Telega, Topics on unilateral contact problems of elasticity and inelasticity, in J.J. Moreau and P.D. Panagiotopoulos, eds., *Nonsmooth Mechanics and Applications*, Springer-Verlag, Wien, 1988, 340–461.
143. R. Temam and A. Miranville, *Mathematical Modeling in Continuum Mechanics*, Cambridge University Press, Cambridge, 2001.
144. C. Truesdell, ed., *Mechanics of Solids, Vol III : Theory of Viscoelasticity, Plasticity, Elastic Waves and Elastic Stability*, Springer-Verlag, Berlin, 1973.
145. H. Tsai and P. Rosakis, On anisotropic compressible materials that can sustain elastodynamic anti-plane shear, *Journal of Elasticity* **35** (1994), 213–222.
146. W.R.D. Wilson, Modeling friction in sheet-metal forming simulation, in *The Integration of Materials, Process and Product Design*, Zabaras et al., eds., Balkema, Rotterdam (1999), 139–147.
147. P. Wriggers, *Computational Contact Mechanics*, Wiley, Chichester, 2002.
148. P. Wriggers and U. Nackenhorst, eds., *Analysis and Simulation of Contact Problems*, Lecture Notes in Applied and Computational Mechanics, Vol 27, Springer, Berlin 2006.
149. P. Wriggers and P.D. Panagiotopoulos, eds., *New Developments in Contact Problems*, Springer-Verlag, Wien, New York, 1999.
150. E. Zeidler, *Nonlinear Functional Analysis and its Applications.* I*: Fixed-point Theorems*, Springer-Verlag, New York, 1985.
151. E. Zeidler, *Nonlinear Functional Analysis and its Applications.* III*: Variational Methods and Optimization*, Springer-Verlag, New York, 1986.
152. E. Zeidler, *Nonlinear Functional Analysis and its Applications.* IV*: Applications to Mathematical Physics*, Springer-Verlag, New York, 1988.
153. E. Zeidler, *Nonlinear Functional Analysis and its Applications,* II/A*: Linear Monotone Operators*, Springer-Verlag, New York, 1990.
154. E. Zeidler, *Nonlinear Functional Analysis and its Applications,* II/B*: Nonlinear Monotone Operators*, Springer-Verlag, New York, 1990.

155. E. Zeidler, *Applied Functional Analysis: Main Principles and Their Applications*, Springer-Verlag, New York, 1995.
156. E. Zeidler, *Applied Functional Analysis: Applications of Mathematical Physics*, Springer-Verlag, New York, 1995.
157. Z.-G. Zhou, B. Wang and S.-Y. Du, Investigation of antiplane shear behavior of two collinear permeable cracks in a piezoelectric material by using the nonlocal theory, *Journal of Applied Mechanics* **69** (2002), 1–3.

Index